A
HISTORY OF
Metals
IN COLONIAL
AMERICA

A
HISTORY OF
Metals
IN
COLONIAL
AMERICA

JAMES A. MULHOLLAND

The University of Alabama Press
University, Alabama

Library of Congress Cataloging in Publication Data
Mulholland, James A. 1935–
A history of metals in colonial America.
Bibliography: p.
Includes index.
1. Metallurgy—United States—History. 2. Metals—
History. I. Title.
TN623.M84   338.973   80–15130
ISBN 0–8173–0052–X
ISBN 0–8173–0053–8 (pbk.)

Contents

Illustrations

Maps

Table

Table 1. Comparative Production of Iron in America

Preface

To one educated in an engineering tradition as I have been, the writing of history is fraught with perils. The adage that facts generated under controlled laboratory conditions and properly presented "speak for themselves" hardly provides a useful guide to the interpretation of historical events, the details of which, in engineering parlance, correctly belong to the realm of nonreproducible data. Rather, the facts of history provide the warp and weft for the historian's loom, and the interpretations thence derived are patterns conceived by study, reflection, and continuous reappraisal. Despite one's best efforts, such patterns rarely remain uncolored by the limitations of one's knowledge or unavoidable bias. In writing history, then, the acclaim for achievement or criticism for weakness in any study inevitably and correctly falls on the shoulders of the author, who stands responsible for the integrity of his work.

No study as broad as the current one can be brought to a satisfactory conclusion without the advice and assistance of others. Here I would like to recognize those whose contributions made this work possible. Among them are George Basalla, John Beer, and George Frick at the University of Delaware and Robert Elliott of North Carolina State University, whose careful reading of the text at its various stages and timely suggestions prevented the perpetration of many an error. My knowledge of the processing of metals was increased by recommendations from Cyril Stanley Smith of the Massachusetts Institute of Technology. A major inspiration for the study was Eugene Ferguson of the University of Delaware, who also served throughout as my "conscience," offering encouragement, criticism, and sympathy at the times and in the proportions needed. And to Malcolm MacDonald of The University of Alabama Press my special thanks for his enthusiastic support of this project.

Among the many who contributed time and effort on my behalf in specific areas I want to thank Carol Hallman of the Eleutherian Mills Library, Greenville, Delaware, Conley Edwards at the Virginia State Library, Michael Musick of the National Archives, Peter Parker and

Lucy Hrivnak at the Historical Society of Pennsylvania, and Lourdes More and Margaret Sugg of the D. H. Hill Library, North Carolina State University, for invaluable assistance in locating and obtaining needed materials. Kathryn Hardee and Ingrid Finnell deserve credit for the preparation of the final manuscript.

Finally, I owe a debt to my wife, Marillyn, without whose understanding and assistance this work could not have been written. As confidant and critic, she shared my frustrations and successes. To her this work is dedicated.

viii

Preface

Foreword

A thriving technology for the smelting and manufacturing of iron and other metals is not normally associated with the colonial period of American history. A general conception prevails among both engineers and historians that the modern role of metals stems from the sweeping changes in the methods and organization of industry in the late eighteenth and early nineteenth centuries, the era identified as the Industrial Revolution. Histories of the metals industries in the United States rarely consider the iron industry significant before its expansion in conjunction with the building of the railroads just before the mid-nineteenth century. Although a few studies, such as those of Arthur Cecil Bining on iron manufacture, have attempted to examine facets of the introduction of metals technologies during the colonial period, the history of that period still is recounted with very little reference to the presence of any technology other than that related to agriculture. In view of the rapid growth of mining, smelting, and manufacture of metals in the early nineteenth century, I felt that a detailed survey of the introduction and early development of metals technology in America could provide an additional dimension for understanding and interpreting the industrialization of the United States. The initial purpose of this study, therefore, was a search for beginnings. It soon was necessary to consider the subject of colonial metals in a broader context.

Voltaire, the master of satire, is purported to have said, "History is after all only a pack of tricks we play on the dead."[1] As a diligent and perceptive observer of the deeds and misdeeds of mankind, Voltaire probably realized that history also is fully capable of playing tricks on the historian. As I set out to chronicle the history of metals in colonial America as a prelude to the study of the role of metals in the industrialization of the nation, a picture began to emerge that gave an altered perspective to the importance of metals in the colonial period. It became increasingly clear that the establishment and development of metals technology in America was essential to successful colonization by the English-speaking peoples and important in fostering the spirit of independence that culminated in the American Revolution and the founding of a new nation.

Although at the time of the settlement of North America by the English-speaking peoples, European society was still agrarian in nature, there were few facets of life that did not involve the use of metal. Agriculture began with the clearing of land with the steel-edged ax, followed by the use of the iron-bladed plow to break the soil. Reaping and threshing with iron implements preceded grinding in a water-driven mill, where the wooden components of wheel, shafts, and gears were fashioned with metal tools and reinforced with iron bands. Crude houses could be constructed from natural materials using no other metal than the ax, but more substantial dwellings required nails and carpentry tools to build and iron, copper, and pewter utensils and hardware to furnish. Transportation also relied on the skills of metals craftsmen, from shaping wagon hardware and the horseshoe to the anchors and chains of sailing ships. On a comparative scale, the consumption of metal per capita has expanded greatly over the last few centuries, but the need for metals in the basic processes essential to civilized existence already had been firmly established before the seventeenth century.

The full significance of metals to an earlier, simpler age becomes more apparent when one considers the problem of attempting to create a society without metals. That significance has been brought home to me in teaching the history of technology and finding that students often manifest a deep mistrust for anything and everything labeled technology. It is an easy jump from the anxiety generated by nuclear arms races, ecological crises, and the erosion of individuality by the computerization of every facet of life to the conclusion that technology is out of control. Inevitably, probing reveals that in the minds of students uncontrolled technology is equated with too much technology. Their idealism conceives the goal of all human endeavor to be the "good life," a utopian society where man at last is free to realize his individual potential. But how does one create such a society? One approach is to ask the students to be master planners, to blueprint a self-sufficient community capable of enjoying the good life free from the pressures and anxieties created by the demands of modern society. Given the option to use or omit any technologies in creating their utopias, one fundamental fact always emerges from the students' efforts: all of the societies, however constituted, require the use of metals as a material base.

Sir Thomas More, the author of *Utopia*, whose work gave the generic name to subsequent speculations on the formation of ideal communities, reached an identical conclusion. Although writing before the industrial era and reflecting the prevailing agrarian basis of English society, he clearly perceived the role of metals in that society. Among the few occupations he listed for Utopians was that of metalworking, and other occupations such as masonry and carpentry required the use of metal tools. Although More endowed his Utopia with independence in nearly all the necessities of life, the one essential commodity specified by name in a brief reference to imports was iron.[2]

If we regard the colonization of North America as the first great effort by Western civilization to create a utopian society, an idea that may warrant further study on its own merits, the colonists can be seen to have confronted in reality the same problems of material supply that concerned More's fictional Utopians. Survival in the wilderness for people nurtured in the habits and culture of Europe required the employment of metals, particularly iron. In the real world of seventeenth-century Europe, the colonists could have imported to satisfy their requirements, exchanging gold and silver for iron. More's Utopia possessed abundant precious metals, although it is often forgotten that Utopia's gold and silver, like its iron, were imported.[3] Tales of great hordes of treasure among the natives of North America and of rich mines were common before and during the colonization period, their lure providing a chimerical magnet for many of the firstcomers. They helped to sustain an interest in finding mineral wealth that yielded only slowly to the more pressing problems of daily existence.

Europe possessed a sufficiently advanced technology to build fleets equipped with cannon capable of waging war on a global scale even before the English came to North America.[4] As long as numbers were few and settlements were located near the coast and along the navigable rivers, that same capability supplied limited needs with relative ease. Thus for several decades the new utopia, like its fictional counterpart, was able to survive on imports. During that period, widespread interest in exploiting American resources was shown by colonists and companies of English merchants, particularly in the southern colonies. The hope was to find precious metals, an extension of the expectation that helped motivate the sixteenth-century voyages of exploration. When gold, silver, and copper were not found, the exploitative urge subsided to be replaced by an entrepreneurial stage in the colonies' development of metals. In the new stage, the knowledge and equipment to build ironworks were sent to the colonies, but the goal was to smelt iron for sale and use in England, not for the benefit of the colonists. Later, the same philosophy was applied to the mining of copper. It was a philosophy shared by English investors and prominent colonial officials and merchants. Like the exploitative stage, the entrepreneurial stage assumed a continuing dependence on England for basic metal manufactures.

History has no laws that define the stage of development in a colonial society, or one emerging from agrarianism, at which the introduction of a particular technology must occur. Neither has history explained the effects of the transformation of a society outwardly dependent for basic material commodities to one independent in the production of those commodities. Whether one degree of material independence fosters a desire for material independence in other areas is a question that lies outside the scope of the present study, although the initial inclination would be to answer such a question in the affirmative. But the degree to which material independence, particularly in such a basic commodity as iron, contributes to a growing sense of political independence must be addressed in considering the development of metals in America. There is strong indication that here also a positive correlation may be found.

As the population grew and the pattern of settlement extended away from the convenient transport routes of the coast, the dependence on English manufactures became strained. The problems of supply were aggravated by political events in Europe and increasing limitations on the fuel necessary to smelt metals in England. The supply of manufactured goods was failing to keep pace with the growing demands for them on both sides of the Atlantic. Those were conditions More did not have to consider in creating his Utopian community. They are the points of departure between the static, ideal community of literary conception and the reality of a dynamic society for which demand and the corresponding problems of supply constantly increase. At some stage in the growth and evolution of society, the very character of need undergoes transformation. What has been an adequate solution to the satisfaction of human wants proves no longer adequate. Need becomes imperative. The alternative to continuing dependence upon technologies employed elsewhere is to import the technologies themselves.

In the early eighteenth century, a growing impatience with the supply of metals from England led to the creation of an indigenous American iron industry. Immigration had brought metals craftsmen to America with their valuable trades, but to build iron furnaces and forges it was often found necessary to import additional skilled workmen as the preferred alternative to importing metal. Amid the shifting fortunes of the early eighteenth-century English-speaking world, those men and their knowledge kindled a colonial desire for material independence that laid the foundations of the metals industries of future generations. Throughout the eighteenth century there was continuous and increasingly strong opposition by Americans toward any efforts by the British government to regulate the indigenous iron industry and metal manufactures. The political friction stemming from the threat to growing material independence represented by British regulatory legislation preceded by several decades the better known crises over stamps and tea. When the culmination of events led to an open break with England, the colonists faced the prospect of fighting a war against an acknowledged military power and the very country on which they had been materially dependent. By persisting in their course of action, the colonials displayed a confidence in their ability to wage war once that dependence had been severed. The existence of a viable metal industry, capable of supplying the materials of war, was a strong factor in inducing that sense of confidence.

The culmination of the revolutionary war ended the colonial status of the United States politically, but how successful had the Americans been in achieving material independence? In an article examining the state of the iron industry during the late eighteenth century, Joseph E. Walker has observed that the end of industrial colonialism did not coincide with the end of political colonialism. American manufacturing, and the iron industry in particular, still was subject to powerful economic restraints which Great Britain could exercise through the creation of trade barriers and the passage of laws to prevent the shipment of machinery or even new knowledge to America. The United States also had to depend on foreign raw material supplies of copper, tin, and zinc for manufacturing and of gold and silver for currency and jewelry well into the nineteenth century. Walker has suggested that for the iron industry the "colonial" period did not end until nearly the middle of the nineteenth century.[5]

For the current study, however, the terminal point chosen has been the decade of the 1790s. At the beginning of that decade, Alexander Hamilton issued the *Report on Manufactures*, strongly recommending the support of manufacturing as a national goal. In words reminiscent of More's conclusions regarding metals in the *Utopia*, Hamilton accorded iron the preeminent place among manufactures. Although actual federal encouragement of manufacturing remained highly variable until well into the nineteenth century, the direction the development of those industries was to take in the nineteenth century already had been delineated by 1800, as Hamilton realized.[6]

There is yet another reason to conclude the present study before 1800. I have attempted to examine, albeit sketchily at some points, the entire spectrum of metals as their mining or manufacture occurred during the colonial period. During the first decades of the nineteenth century, rapid expansion occurred in existing metals industries and new ones were added. The iron industry continued to grow along the eastern seaboard, particularly in the Pennsylvania-Maryland region, and it also accelerated the movement across the Appalachians begun in the 1790s. There were new developments in the mining and manufacturing of copper and lead, the beginnings of American brass and tin industries, new explorations and resultant mineral finds stimulated by the Louisiana Purchase and the westward movement of population. Finally, there was the discovery of gold in western North Carolina, the fulfillment of a quest begun before the colonization of North America. All but the last event more properly belong to a new era, to a continuous and accelerating growth of technology that transformed the United States from an agrarian to an industrialized nation. Hence the decade of the 1790s represents a watershed in the history of metals in America.

The story of the introduction and growth of the technology of metals in the colonial period, then, entails significant developments beyond the transfer of the technology from the Old World to the New. In the determined struggle to create an indigenous industry, in the efforts to encourage and support the work of metals craftsmen, in the colonial defiance of British attempts to regulate manufacturing of metals, the importance of metals to colonial society is unmistakable. The establishment of metals technology, the basis for its future industrial greatness, was firmly intertwined with the political foundations of the United States in the colonial period.

Foreword

A
HISTORY OF
Metals
IN COLONIAL
AMERICA

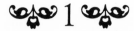

Before Jamestown

Men can no more do without iron than without fire and water. But gold and silver have no indispensable qualities. Human folly has made them precious only because of their scarcity.—Sir Thomas More, Utopia, 1516[1]

Seagull: Come boys, Virginia longs till we share the rest of her maiden-head.

Scapethrift: But is there such treasure there, captain, as I have heard?

Seagull: I tell thee gold is more plentiful there than copper is with us; and for as much red copper as I can bring, I'll have thrice the weight in gold. Why, man, all their dripping-pans and their chamber-pots are pure gold, and all the chains with which they chain up their streets are mossy gold, all prisoners they take are fettered in gold; and for rubies and diamonds, they go forth on holidays and gather 'em by the seashore to hang on their children's coats and stick in their caps, as commonly as our children wear saffron-gilt brooches and groats with holes in 'em.—George Chapman, "Eastward Ho," act 3, scene 3, 1605[2]

When Sir Thomas More wrote his *Utopia* at the beginning of the sixteenth century, it was with the full and certain knowledge that iron supplied the material backbone of Western civilization. A new world recently had been discovered beyond the Atlantic. On the islands of the Caribbean and along the coasts of Central America the early explorers encountered relatively primitive tribes, uncontaminated by the institutions of European civilization, but also lacking any evidence of the technical skills and material culture deemed essential by Western man. More, intent on satirizing the growing corruption he perceived in his native England, seized upon the as yet uncharted expanses of America for the site of an idealized community. Certainly, it could be argued, the reports of explorers attesting to the unsophistication of the natives afforded a telling contrast to the social and moral degeneration that many felt characterized Europe. Nevertheless, More believed, the concept of an advanced civilization without metals would have been incomprehensible to any knowledgeable European. Because America in 1512 appeared to lack both metals and the skills to work them, he felt impelled to find some means to supply his Utopia with both. The acquisition of Western technological skills More attributed to knowledge gained from a body of Roman and Egyptian artisans shipwrecked twelve hundred years prior to the pretended narrative. More circumvented the absence of metals by having the Utopians import them— iron for basic uses and, a point often overlooked by later seekers for utopias in America, gold and silver for commerce. As the source of much of the misery existing in contemporary society, the Utopians held gold and silver up to scorn and ridicule. "They have," More noted, "no indispensable qualities," whereas "iron is already greatly superior to either." More's observation was a sound reflection on recent developments in the use of metals throughout Europe.[3]

Since the late Middle Ages, Europe had been experiencing a steady growth in industry and trade. Expanding industries, increasing uses of power, the revival of cities, and changes in the technologies of farming and warfare all had made new demands upon the supply of raw materials. Widespread trade, reaching as far as the territories of the great Khan, encouraged an increase in the supply of precious metals. Mines abandoned since the period of the Late Roman Empire were reopened, and an active search for ores in hitherto little-explored regions gave rise to flourishing mining districts in Saxony, Thuringia, Hungary, and Sweden. In areas such as Devon and Alsace, where older mines existed, new veins of silver-bearing ores were discovered. Throughout Europe the production of metals, particularly iron, steadily increased.[4]

As the metal of paramount utility in Western civilization, iron's abundant ores were recognized and exploited in every country. Iron, however, was difficult to recover. The high temperatures necessary to effect complete separation of the metal from its oxides were beyond the technical capacities of the furnaces used in Europe before the introduction of the mechanically powered blast in the fourteenth century. The evolution and spread of the blast furnace made iron more readily available with a corresponding increase in its use. The metal could be worked easily only when red hot and was subject to mysterious maladies causing some pieces to crumble under the hammer when hot and others to crack when cold worked. Moreover, the role of carbon in making steel was not understood before the late eighteenth century.[5] For all its widespread use and utility, the smelting and working of iron constituted one of the most esoteric technologies discovered by Western man. As More observed, however, preoccupation with mining gold and silver from the twelfth through the sixteenth centuries often obscured the importance of iron.

Although far less common, the noble triad of gold, silver, and copper are relatively inert chemically so that they can be found in the pure or "native" state in nature. It had long been known that silver often occurred in association with lead ores, and techniques for the recovery of the small quantities of silver in lead ore had long been developed. Copper can be reduced from its ores at relatively low temperatures, the same being true of the baser metals tin and lead. Together with gold and silver those metals can be worked extensively when cold. But the relative scarcity of the noble metals meant that far fewer people were familiar with their ores than with those of iron; hence, the ability to locate profitable lodes where native metal or lead ore outcroppings were not conspicuous generally was confined to the inhabitants of the active mining districts, most notably the German states. Given the European climate that fostered an awareness of the importance of metals, it was inevitable that the search for their ores would be carried to any lands beyond the Atlantic, but the emphasis was on the scarce metals, gold, silver, and copper.

Long before 1605, when George Chapman parodied More's words (Seagull's speech above is a direct recital of the passage describing the Utopians' contempt for gold), gold and silver had become a magnet drawing the Europeans westward. Spain reaped the benefits of early successes, inspiring a century of exploration by the European nations and invariable failure to discover new sources of precious metals. Rumor, however, proved more powerful than sober fact. The voyages of discovery—the quest for gold, silver, and copper in North America—set the stage for its colonization. Meanwhile, the lesson preached by More concerning the fundamental importance of iron was forgotten. During the sixteenth century, no one saw the necessity to relearn it.

Sir Thomas More still was twenty years away from writing *Utopia* when Columbus discovered America. In 1492, after fruitlessly offering his services to several European monarchs, including Henry VII of England, Christopher Columbus sailed westward under the banner of Spain to seek a new route to the Orient. Instead of a shortcut to the Indies, he found a new world, although at first he did not realize it. On subsequent voyages, the Spanish under Columbus planted the first European colonies in the Western Hemisphere since the Norsemen, and from them expeditions fanned out to the north and south, partly in the hope of finding treasure, partly still in search of a route to the East. In so doing, they established a pattern other nations, particularly the British, would follow in developing their interests in North America.

The coastal tribes first encountered by European explorers had little knowledge of metals, their sources, and uses; but, as Sir Thomas More had envisioned in *Utopia*, there were civilizations in the Americas that had extensive experience in the recovery and working of at least the more easily discoverable noble metals. Hernando Cortes's expedition to the central valley of Mexico in 1521 first brought Europeans into contact with them.

In the highly developed civilization of the Aztecs, and again in the Incan Empire in South America, the Spaniards encountered people skillful in the working of metals but with different attitudes toward metals from their own. Throughout Mexico and Central America gold was recovered from alluvial deposits. Silver and copper were obtained in the native form, usually from surface deposits or outcroppings, although the Aztecs had begun to develop underground mines for silver-bearing ore. As surviving examples of jewelry and plate objects illustrate, the Aztecs were capable of fashioning elaborate works of art and ornamentation from those metals.[6]

Incan metallurgy was the most highly developed in the Americas, in some instances on a par with the finest European practice. The Incas were masters of the smelting and casting of gold and silver. Among their techniques was the sophisticated "lost wax" process in which objects with complex features can be modeled in wax. A mold then is built around the model and heated, allowing the wax to melt and run out, leaving a void in the shape of the original model. Molten metal is poured into the mold, which may have a sand core, to produce, when finished, a hollow object with fine surface detail. The Incas used the process to cast gold and silver into intricate and beautiful shapes. Shortly before the Spaniards arrived, the natives appear to have discovered the tin of Colombia and used it with copper to make bronze. Most of their work was decorative or ornamental (like that of the Aztecs), but some practical applications of metals already had evolved, such as the use of copper for spear points or surgical tools. The Incas were adept at welding and gilding and had discovered means of working with the metal platinum. Platinum, found in alluvial deposits in the Andean region, was unknown in Europe before the conquest of the Americas, and its extremely high melting point proved a barrier to its use by Europeans for many years. The great number of artifacts produced by the Incas incorporating platinum testify to their metallurgical skill.[7]

The Amerindian civilizations did not use iron, nor did they know how to prospect for, mine, or smelt it. Both the Aztecs and the Incas ruled extensive empires, secured by military conquest and maintained by superior force, but conquest had been achieved without metal weapons. They did possess gold, silver, and copper, in staggering abundance to the eyes of the conquistadores, but the role to which metals were relegated was, except for the absence of iron, strangely reminiscent of *Utopia*. They wore gold on their arms and breasts, dangled it from their ears and noses, hung it on their walls, made it into statues to decorate their gardens, buried it with their dead. But they did not regard gold and other metals as personal wealth in the Western sense.

When the civilizations of the Old and New Worlds clashed, to a large extent it was because of metal—the high value the Europeans placed on the gold and silver they found among the tribes of the Americas. Yet in the struggle, More's words regarding the significance of iron acquired new meaning. If the struggle was over the possession of precious metals, the outcome was determined by iron. The superiority of the Spaniards rested in their cannon, swords, and armor, and with these weapons the iron-based civilization emerged triumphant. Although Western man coveted gold and silver, it was by the use of iron that he was able to establish hegemony over the New World. The Spanish, however, in seeking to exploit the fruits of their conquests, made no immediate effort to mine and manufacture iron in America. Their monarch, Charles I, also wore the crown of the Holy Roman Empire, and his hereditary lands included the rich silver-mining districts of Austria and Bohemia. Borrowing the techniques of the German miners, the Spanish sought out and developed the sources of Aztec and Incan silver. Their first mine in Mexico was opened in 1522, the year after Cortes's victory. Within a few years annual shipments of silver from mines at Zacatecas in Mexico and Potosí in Peru crossed the Atlantic to swell the treasury of Spain.[8]

The Spanish conquest of the Aztec and Incan civilizations with their treasures of gold and silver was the decisive factor in establishing European attitudes toward the potential of the New World. The traditional methods for obtaining metallic wealth usually required exhaustive searches for traces of ores and distinct elements of risk where supplies had to be assembled and exploratory shafts dug to open up veins that might prove worthless. There was considerable drudgery and danger in the darkness and dampness of the mines and in the heat of the furnaces. All of these elements were lacking in the early stages of the Spanish adventure. In America, it seemed, as in More's *Utopia,* the streets were chained with gold, ready to be taken. Success had come too easily, and human folly was transcendent. As yet no one really knew the extent of the Americas; a vast area remained to be explored. No one knew or could even guess at the treasures that still might be waiting to be found. As the bullion fleets moved eastward to Spain year after year, the impression developed that endless wealth still lay untouched beyond the western shores. Spain's achievements in the first half of the sixteenth century drew the attention of her rivals, France and England, to America and established the climate of anticipation for the northern voyages of exploration, a climate in which the search for gold and silver came to overshadow other goals and, often, men's reason.

It was an accident of geography that brought Columbus to the Caribbean threshold of the metal-rich civilizations of the Americas. Had Columbus struck northward to the lonely beaches and barrier islands of the Carolinas or had he made landfall along the jungled coasts of the Amazon region, a different chapter in Western history might have been written. The early English experience in the waters of the New World offers marked contrast to that of the Spanish adventurers.

England had made its first sorties into the waters of North America well before the conquistadores discovered the central valley of Mexico and the rich civilization of the Aztecs. At that time, the young Thomas More still pursued his law studies at Lincoln's Inn, preparing for his career as a public servant. When, in 1496, King Henry VII granted a royal patent to the Genoese expatriot, John Cabot, for a voyage of exploration, it was mainly with the expectation of finding a new route to the Indies. England's trade was dominated by foreigners, and her preoccupation with the sea, later to become a national pride, had languished after decades of foreign and domestic conflict. Henry's earlier reluctance to sponsor Columbus probably emphasized as much the dearth of experienced seamen necessary to support so ambitious a venture as the inadequacy of a royal treasury recently depleted by the long civil Wars of the Roses.[9] By all accounts, Henry was a prudent man when it came to money, one who certainly was aware of the value of commerce, and Columbus's mistaken report of having reached the Indies by sailing westward offered a prospect for the revival of British trade. It is perhaps characteristic that Henry's first patent to Cabot did not specify commodity, just the royal share. The king agreed to accept one-fifth of any profits to be realized while offering his blessing (but no financial support) in return.[10]

The exact details of Cabot's voyage and of his discoveries have been a subject of speculation for historians because no log has ever been found. It is known that John Cabot left Bristol in May 1497, returning there in August of the same year after having discovered lands somewhere far to the west of Ireland, most probably in the vicinity of the Canadian Maritimes. Contemporary accounts agree that Cabot discovered a "newfound land" with rich fishing waters, but there is no direct evidence that he found either metals or ores in his travels. Cabot, like Columbus, was convinced that he had arrived at the borders of the territories of the great Khan, and Henry was sufficiently satisfied to grant Cabot a small pension for his efforts. The pension was paid only briefly; Cabot disappeared with a number of English merchants and four ships on a second expedition into the Western Ocean in 1498.[11]

Cabot's two voyages at the end of the fifteenth century, and one by his son Sebastian who sought a passage to the Orient in the sixteenth by sailing northeastward from England into seas of ice, seem to have stirred far less interest in England at large than did those of the Spanish and Portuguese mariners in their countries. The king's total outlay to support the voyages probably was less than £300, and that was mostly in the form of gifts and pensions to the two Cabots. The returns were even less.[12] Henry subsequently granted letters of patent to several English merchants encouraging the importation of gold and silver and other commodities from western lands, but nothing ever came of these. The first efforts of the English to explore North America ended without their finding a passage to India. Nor did they discover spices or gold or any other product of value.

It is not surprising, then, that when Henry VIII ascended the throne in 1509 he evidenced no interest in the New World that had shown so little promise. Turning his attention to the Continent, he soon became embroiled in political quarrels and a series of costly wars, dissipating the large treasury amassed by his father. In these ventures Sir Thomas More, Henry's chancellor, had opportunity to witness how metals, and in one important instance the lack of them, affected the king's designs and his country's well-being.[13]

The early sixteenth century saw an acceleration of changes in the nature of warfare, particularly in the use of field artillery, an innovation first used by the French under Charles VIII at the Battle of Fornovo in 1495.[14] The new weapon initially was made of cast bronze, an alloy of copper and tin, using essentially the same techniques developed and perfected in an earlier age to cast cathedral bells and doors. New uses also were found for iron in hardware for gun carriages, in the manufacture of muskets, and in the building and outfitting of Europe's expanding navies.

England possessed ample deposits of iron ore, and in the fifteenth century sufficient bloomeries existed to supply the normal demand for ironware. Moreover, the blast furnace had been introduced into the British Isles from the Continent between 1490 and 1500. The development of the Weald of Sussex as a major iron-producing region followed shortly after. Yet, despite both improvements in iron technology and the spread of the industry into new areas, the price of bar iron rose from £3 per ton in 1490 to £9 in 1547, as the greater utility of iron encouraged ever greater demands for the metal.[15]

Bronze remained the preferred material for cannon, both on ships and in field artillery. Because it melted at a lower temperature than iron, sound castings could be obtained consistently. Bronze was easier to bore to obtain a uniform diameter of the barrel with the crude tools then available, and it did not corrode when exposed to rain or salt air. Although England possessed abundant tin in the ancient mines of Cornwall, equally ancient copper mines had been abandoned for centuries, so that copper was not thought to be found in England. The principal European sources were in Sweden and in what is now southern Germany. Early in his reign, Henry attempted to encourage the discovery and development of mines in England, particularly for copper, gold, and silver. Henry's brother-in-law, Charles Brandon, duke of Suffolk, organized the attempt on the king's behalf. The efforts failed, however, and one of the reasons Suffolk noted was "unskillfulness in his labours, the cunning of such kind of work being not then thoroughly known."[16] To wage war in the "modern" style, therefore, the king was forced to place large orders abroad for the casting of ordnance, producing a substantial drain on England's coffers.

English interest in the New World waned while Henry occupied himself in Europe, except for one brief spate of activity. In 1527, the London merchant, Robert Thorne, then a resident of Seville in Spain, wrote a letter to his Most Excellent Prince, exhorting him to undertake the exploration of the northern waters to seek a sea route to the Indies that he might "with little cost, perill, or labour . . . amplifie and inrich this your sayd Realm." Visibly impressed by the success of Portuguese trade with Africa and India and with the initial reports of the gold and spices available to the Spanish in the newly discovered lands to the west, Thorne's enthusiasm was unrestrained. He declared that between the equinoxes were to be found the richest lands and islands in the world. Instead of English apples, nuts, and corn, they yielded peppers, cloves, dates, and spices, and where "our metalls be Lead, Tinne and Iron, so theirs be Gold, Silver and Copper." Portugal and Spain had vastly enriched themselves by sailing west, south, and east. Thorne implored his king "that now rest to be discovered the sayd North ports, the which it seemeth to mee, is onely your charge and duety."[17] Henry, frustrated in his desire to find precious metals at home, responded to the hope of finding them abroad by supporting another voyage of exploration. Subsequently, John Rut sailed westward in 1527 with the king's blessing and blundered along the northeast coast of America for some time before finally returning home via the West Indies, having contributed little new to knowledge of the region and nothing to Henry's pocketbook. Rut's profitless voyage brought a long halt to English interest in the New World. Beset by religious turmoil in the wake of Henry's break with the church in Rome, financially depleted by the king's profitless wars and extravagant ways, England had little energy to spend exploring the unknown lands across the Atlantic or strength to intrude upon those already occupied by powerful Spain.

In 1524, France, represented by Giovanni di Verrazzano, joined England and Spain in the ranks of countries exploring the New World. The Florentine captain, financially backed by countrymen who were bankers in Lyons and Rouen, set sail in a borrowed French naval ship in January 1524. His primary goal was clear—to find a western passage to Cathay in the unknown seas north of the Spanish-controlled lands, the same goal Thorne urged upon Henry a few years later. From Columbus to Magellan, the Spaniards had explored the coasts of the Americas, starting with the Caribbean and moving southward. They found an unbroken expanse of land all the way to Tierra del Fuego, where Magellan's fleet finally was able to turn westward into the Pacific en route to the first circumnavigation of the globe. North of Florida, however, there was only Cabot's newfound land, but whether that was an island or a continent no one knew.

France already had found a way to tap the treasures of America by capturing homeward bound Spanish ships with their rich cargoes. Thus Verrazzano, his French flag unwelcome in Spanish waters, elected to steer well north of the route taken by Columbus. Consequently, he made landfall at what today is Cape Fear, North Carolina. From there, Verrazzano sailed northward along the coast as far as Newfoundland without finding a westward passage. One incident of his journey, however, is of significance to the present story. He entered Narragansett Bay in Rhode Island where he found the Wampanoag Indians. He made special mention of their "plates of wrought copper, which they esteem more than gold," native copper that had been obtained from the Great Lakes region through trading.[18] The copper so eagerly desired by England had first been noted by her old enemy, but the source of the copper long eluded discovery.

Following Verrazzano, the French made a series of voyages to explore the New World. A major factor in the French decision was the declaration in 1533 by Pope Clement VII that the bull of Pope Alexander VI in 1494 dividing the Americas between Portugal and Spain applied only to lands already discovered. New lands and any treasures therein would belong to the finder.[19] Almost at once, in 1534, King Francis I dispatched two ships under the command of Jacques Cartier to discover, in the name of France, new islands and countries "said to contain gold and riches." The voyage produced the first substantive knowledge of the geography of the Canadian maritime region (Cabot had been extremely vague on this), but little of potential value except to reaffirm Cabot's discovery of the teeming fishing grounds offshore. On a second voyage, Cartier sailed far up the St. Lawrence River in search of a mythical kingdom called Saguenay, somewhere to the west, reported by the natives to be abounding in precious metals. At the site of Montreal, he came across Hurons wearing copper ornaments which he was told also came from western lands. Cartier returned to France for the second time with a detailed and realistic account of the country and its resources colored only by a persistent belief in the existence of Saguenay.[20]

The sixteenth century was a time of fabulous deeds, and the legend of an earthly paradise first nurtured in the Middle Ages was revived and recounted throughout Europe. Why should not it and its riches await discovery in the unexplored reaches of the New World? More's fictional Utopia was one manifestation of this dream. Coupled with the revelation of the Aztec and Incan civilizations, such thinking helped to create and sustain the reported myths of richer kingdoms hidden somewhere to the west. Like the Seven Cities of Cibola, which drew the Spanish in the person of Coronado north from Mexico as far as the plains of present-day Kansas, and the lure of El Dorado in the Guianas, which later attracted Sir Walter Raleigh and the English, the mythical wealth of Saguenay tantalized the French. Buoyed by the accounts of Cartier's initial explorations, the French king authorized another expedition.[21]

The third voyage of Jacques Cartier was to be more than an exploration; he fully intended to establish a colony. In anticipation of the discovery of the kingdom of Saguenay, the preliminary list of settlers included blacksmiths, metallurgists, and jewelers. When he actually sailed in 1541, however, Cartier's "colonists" consisted of the all-too-common collection of convicts and other riffraff.[22] After landing at a point near Cape Rouge, they began the work of settlement in earnest, as well as a search for gold. Richard Hakluyt's translation of the official French account related: "And upon that high cliffe wee found a faire fountaine very neere the sayd Fort: adjoyning where unto we found good store of stones, which we esteemed to be Diamants. On the other side of the said mountaine and at the foote thereof, which is towards the great River is all along a goodly Myne of the best Yron in the world, and it reacheth even hard unto our Fort, and the sand which we tread on is perfect refined Myne, ready to be put into the fornace. And on the waters side we found certaine leaves of fine gold as thicke as a man's nayle."[23]

Cartier dispatched two ships bearing samples of his jewels and gold. Though well received at first, the former turned out to be only worthless quartz and the latter iron pyrites (fool's gold). The ore samples, the only tangible results of Cartier's extensive explorations, were dumped into the Seine. The French abandoned the search for the gold of Saguenay, but the myth persisted as a part of a growing lore concerning great wealth yet to be discovered in the Americas. The lure of gold and silver continued to cloud European perceptions of the New World.[24]

The termination of Cartier's third expedition coincided with changes occurring among the nations of western Europe, and for a few decades further interest in the exploration of North America faded. The already substantial flow of gold into Spain was augmented by a flood of silver from the mines of Mexico and Peru. Instead of enriching the crown, however, the new wealth seemed only to stimulate new wars and other costly ventures. First Charles I and then his son, Philip II, began to borrow heavily to support military campaigns, the loans being construed as advances against the American treasure. The balance of power tilted heavily in Spain's favor and was regarded with growing alarm by the other nations of Europe. One example illustrates the magnitude of the problem. In the autumn of 1566, the arrival of the silver fleet at Seville with a record cargo enabled Philip to equip an army of nine thousand men under the duke of Alba, who marched northward. Alba's force struck terror among European Protestants, who were at first uncertain of his goal, which was to quell insurrection and restore Catholic orthodoxy to the troubled Hapsburg possessions in the Netherlands.[25]

Spanish wealth was used for more than just the overt accoutrements of war. As William Camden later observed, "The Spaniard carried on his Wars not so much by the Strength of Spain as with the Gold of America, whereby he every-where dived into the Secrets of Princes, corrupted and discovered their Counsels, and undermined their Subjects Fidelities."[26] The other nations of Europe had no such resources with which to counter Spanish militarism. But the annual arrivals of the silver fleets had another effect on Spain and her neighbors. The influx of specie touched off a hard inflation that gripped the Spanish economy and sent disruptive waves throughout the rest of Europe. At its peak flow in the last decade of the sixteenth century, the gold and silver imported into Spain reached seven million pesos, but the expenses of wars, including the costly and vain building of the great Armada of 1588, had outstripped income long before that. Spain had sailed westward hoping to find new sources of wealth and had succeeded only too well. The ensuing developments bankrupted the country.[27] The other nations of Europe did not profit from Spain's example.

In the decade after the issuance of the Bull *Regnans in excelsis* in 1570 by Pope Pius V excommunicating Elizabeth from the Catholic church and declaring her throne vacant, relations between England and staunchly Catholic Spain deteriorated.[28] At the same time, England was enjoying a new surge of prosperity. Possessing to some degree the financial prudence of her grandfather, Henry VII, the queen strove to avoid costly foreign entanglements while promoting peace at home. She also succeeded in one area where her father's efforts had failed. In 1568, the Company of Mines Royal was created to mine copper in England. Her father's efforts showed that the "cunning of such kind of work" was not familiar to the English; Elizabeth therefore employed experienced German copper miners. They made the discoveries on which the mines were based, and Germans continued to be instrumental in the commercial and technical development of the mines during the company's existence (1568–1605).[29] A measure of Elizabeth's success was a renewed interest in the promotion of commerce by English merchants. Confronted with Spanish dominance of the seas to the south and the hazards of trading with the Continent in the face of the protracted Dutch-Spanish conflict, the merchants, seeking to expand their limited opportunities, reopened the vision of the Northwest Passage.

In 1576, a company was organized to conduct a new search in the uncharted waters north of Labrador. William Camden credited the origins of the Frobisher expedition to "some studious heads, moved with a commendable desire to discover the more remote Regions of the World and the Secrets of the Ocean." Modern studies tend to agree that profit rather than knowledge supplied the chief motivation. As their captain, they chose a thirty-seven-year-old adventurer, Martin Frobisher.[30] Up to the time of his first voyage to America, Frobisher seems to have been involved in a number of ventures of doubtful legality. The aborted search for the Northwest Passage was to bring him fame and respectability; when, in 1594, he died of wounds sustained in battle, he was a national hero, veteran of the Armada and the Channel engagements against the French. On that first voyage, however, he found only ice and Frobisher Bay, north of Hudson Bay. From Hall's Island, at its mouth, Frobisher also brought back a heavy, black, metallic-looking stone.[31]

Frobisher's stone was analyzed by several competent men and reported to be only iron pyrites. But Michael Lok, a strong supporter of the expedition, persisted in seeking new opinions until he received from one John Babtista Agnello, an Italian assayer living in London, the assurance that the ore contained gold. The obvious question as to how Agnello could find gold where others discerned only worthless rock never has been answered satisfactorily.[32] Although the report was highly questionable, the will to believe was stronger than prudence. In 1577, the Cathay Company was chartered with Lok as governor for life, and the queen herself subscribed a third of the paid-up capital of £3,000. The search for the Northwest Passage was pushed far into the background. Frobisher was dispatched to Hall's Island with three ships and a contingent of Cornish miners to find more ore and bring it back as soon as possible. Two hundred tons of ore were returned to England; part was sent to Bristol, the rest to safekeeping in the Tower of London. German assayers employed by the company declared that the ore would yield gold and silver to the profit of £5 a ton. The excitement of this news touched off another wave of investment, and Frobisher sailed again in 1578 with a larger fleet and more men.[33]

On the third and last voyage, "ore" was discovered at a number of sites in the vicinity of Frobisher Bay, and another large quantity was shipped back to England. Efforts to reclaim the supposed wealth the rocks contained went on for several years. Two assays of the Frobisher ore were conducted by an English assayer, William Williams, using the lead extraction method. In this process, lead forms an amalgam with gold or silver, separating it from the ore. The lead is removed by oxidation. His results were listed in a letter dated July 28, 1583. Using 1 cwt. of Frobisher's ore and 2 cwt. of lead ore, and then using 1 cwt. of Frobisher's ore and 4 cwt. of lead ore, Williams found quantities of silver smaller than a pin head. The two silver samples were stuck to the letter with sealing wax.[34] Perhaps not surprisingly, Frobisher was blamed for having brought back worthless rock instead of the "rich" ore he discovered on his first voyage. Finally, as William Camden noted, "which Stones, when neither Gold or Silver nor any other Metal could be extracted from them, we have seen cast forth to mend the Highways."[35] The whole affair throws into question the state of the arts of prospecting and assaying in England in the sixteenth century.

The English inability correctly to identify metal-bearing rock, other than the ores of iron, presents one of the curious mysteries associated with their exploration of America. The problem was to persist throughout much of the colonial period. Basically, the finding of ores as the preliminary to a mining operation is a two-step process. First is the prospecting step, the detection of likely looking ore bodies or outcroppings. Then comes the assaying step, the evaluation of the quality of the ore, estimating its potential yield by chemical analysis.

The techniques of assaying ores were well developed in Europe by the sixteenth century. To centuries-old techniques of the separation of metals from their ores had been added analytical methods, particularly the use of the balance, developed by alchemists from the Middle Ages onward. The invention and spread of printing stimulated the writing of technical treatises, and in those devoted to metallurgical arts assaying was accorded a prominent place. As Vanoccio Biringuccio wrote in his *Pirotechnia*, first published in Venice in 1540, "Once [ores] are found . . . it is necessary to assay them in order to know what is in them, since the judgment of the sight is not sufficient for recognizing the quantity or even the substance they contain. For this reason, it is necessary to come to assaying and, with the knowledge of experience, to weigh their virtues and to proceed with the work or withdraw from it and its expense."[36] Written in Italian, the *Pirotechnia* with its fund of practical information went through four editions in twenty years, but Cyril Stanley Smith has discovered that no English translation seems to have been made before the nineteenth century.[37] It is, however, entirely probable that when Agnello made his assay of the ore brought back by Frobisher, Biringuccio provided his handbook.

The most famous of the sixteenth-century metallurgical treatises was the *De Re Metallica* by George Agricola, first published in 1556. There already existed a number of assaying manuals for use in the extensive mining districts throughout Germany, but Agricola's work far surpassed them in detail and treatment of all phases of mining and smelting, including assaying methods. Written in Latin, it was available for use to a far wider audience than the *Pirotechnia* or the German manuals. Once again, however, no English translation existed until Herbert and Lou Hoover produced their definitive version early in the twentieth century.[38]

The absence of an English literature does not, by itself, explain their weakness in finding and evaluating ores. As the results of Cartier's expeditions show, such weakness was not confined to the English. The fact remains, however, that the knowledge and skills necessary to prospect for and evaluate ores had yet to be mastered by the English-speaking peoples.

Although England still had received no direct profit from her voyages of exploration, like France at an earlier date she began to achieve access to the metallic wealth of the Americas through seizure of Spanish vessels homeward bound from Mexico or Panama. Sir Francis Drake's voyage around the world (1577–80) touched off a storm of controversy when the Spanish government demanded restitution for the treasure he had seized while coasting the Americas. Elizabeth's response was to deny Spain's proprietary rights in lands that were not already occupied. Henceforth, England claimed the sanction of the law of nations for any country to transport colonies to those uninhabited areas.[39] Nor was the first English colonizing attempt long in coming. In 1582, Sir Humphrey Gilbert, who had acted as a propagandist for Frobisher's voyages, organized an expedition. He set out the following year. Among his goals was to find yet another legendary kingdom in the Americas, Norumbega.

Whatever the origins of the myth, which appears to be purely of European fabrication, the name Norumbega began to appear on maps of North America as early as 1569, when Mercator used it to designate both a region and a city in the Penobscot country in Maine. The word itself is of Algonquin origin, "quiet place between two waters."[40] The myth exploded into prominence in England in the early 1570s, based on the fanciful tales of an English sailor named David Ingram, who somehow had achieved the feat of walking through Indian country from the Gulf of Mexico to the St. John's River in New Brunswick. When examined in a private inquiry into the possibilities for exploration and settlement in America, in 1582, Ingram described a country abounding in gold, silver, and pearls, where nuggets as big as his fist lay in small running brooks and springs. There were, he reported, great cities, whose inhabitants wore on their arms and legs hoops of gold and silver garnished with pearls. Their houses were supported by pillars of gold, silver, and crystal, and, shades of *Utopia*, the scoops, buckets, and vessels of common use were all of massy silver. Gilbert recorded the story as Ingram related it to him, and Richard Hakluyt duly included it in an early edition of his *Voyages*.[41]

The riches of Norumbega beckoned Gilbert, but previous English experience suggested prudence. Even the Spanish had not enjoyed instant success among the rich kingdoms of the tropics. Gilbert, therefore, laid careful plans for the creation of a permanent colony enjoying the "lands, liberties, freedoms, immunities and commodities" due to all Englishmen. Gilbert's proposed plan for the colony had definite utopian undertones. From the colony, the search for Norumbega could proceed without haste. An elaborate and exclusive agreement for trade to and from the colony was arranged with a company of merchant adventurers representing the town of Southampton. His backers included a broad spectrum of English society ranging from merchants and yeomen of Southampton, to Sir Thomas Bromley, lord chancellor, and the earl of Sussex, high chamberlain of England. In June 1583, the company, numbering about 260 men and including "shipwrights, masons, smiths, mineral men and refiners," set sail in five ships for America.[42]

The Gilbert expedition was plagued with bad luck from the beginning. After taking formal possession of Newfoundland in the name of the crown and collecting a store of minerals believed to contain silver, Gilbert and his company sailed south to Sable Island to replace their provisions. The *Delight*, bearing the precious ore samples, was lost in passage together with most of the crew, including the "Saxon" assayer, Daniel, who had conducted the search for ores. Foul weather, lack of supplies, and the loss of their companions discouraged the rest from proceeding to establish a colony during that season. It was resolved to return to England and to make another attempt the following year. Off the Azores, the two remaining ships encountered a fierce storm. Toward the evening of September 9, Gilbert was seen sitting near the stern of the smaller ship, in his hands a book, most likely Sir Thomas More's *Utopia*. More's picture of an ideal society had been one of the inspirations, Norumbega another, for Gilbert's desire to found a colony

in the New World. During the night, his ship and its entire complement vanished beneath the waves of the storm-tossed Atlantic.[43]

Soon after Gilbert's untimely end, his half-brother, Sir Walter
Raleigh, obtained an extension of the Gilbert patent and began to plan
for colonies in the region just north of Spanish Florida. Raleigh's
intentions were no secret. In 1584, Richard Hakluyt of Oxford had
written the "Discourse of Western Planting" at Raleigh's request, in
which the provision of bases from which to intercept the annual
treasure fleets was given as one important reason for establishing
colonies in America. "For," wrote Hakluyt, "take away his [Philip of
Spain's] treasure, . . . his olde bandes of souldiers will soone be
dissolved, his purposes defeated, his power and strengthe diminished,
his pride abated, and his tyranie utterly suppressed."[44]

Raleigh, then, hoped to use the colony as a base of operations from
which to attack and capture the Spanish treasure fleets that moved
northward along the Florida coast on their annual voyages to Spain.
Like Gilbert, he reasoned that the colony might be a starting point
from which the land to the west could be explored for gold and silver.
Possessing less vision than Gilbert, however, Raleigh was primarily
concerned with immediate material gain. When the Roanoke colony
failed in 1589 after a brief existence, he sold his rights in Virginia to a
company of London merchants, while reserving to himself one-fifth of
all the gold and silver subsequently discovered there. Shortly afterward,
Raleigh's attentions were directed to the Orinoco region of South
America, where the fabulous treasure of El Dorado supposedly waited.[45]
In all respects, Sir Walter Raleigh typified the gold-fever mentality that
dominated English perceptions of the New World throughout the
sixteenth century. Although his colony left no lasting imprint in
America, one incident relevant to a history of metals warrants notice.

On a precolonizing expedition of 1585–86, Raleigh sent young
Thomas Hariot as a scientific observer to explore and to report on the
findings of potential assets and on the nature and customs of the
natives. Among the more valuable products Hariot listed in the
Virginia region were copper and some silver worn by the natives and an
abundance of iron ore that he felt "may be allowed for a good
merchantable commodity, considering there the small charge for the
labour and feeding of men, the infinite store of wood, the want of wood
and deerenesse thereof in England, and the necessity of ballasting of
ships."[46] It was a perceptive observation, an echo from the pages of
Thomas More, although it created little or no excitement at the time.

Rather than Raleigh, Sir Humphrey Gilbert had most nearly united
in one person the incentives that drew the English to the American
shores. Throughout the sixteenth century, the search for gold, characterized by the willing acceptance of such myths as Norumbega, had
predominated. To this, Gilbert added a second, the conception of a
colony cast in a utopian mold, a harmonious society free from the
dissension and constant warfare that characterized much of Europe in

the aftermath of the Reformation. Seagull's mocking message quoted at the beginning of this chapter showed how strongly the first theme persisted in the English mind on the eve of colonization. Within a few years, other groups of Englishmen, more secure in the knowledge gained by the efforts of the sixteenth century and undeterred by the failures, looked to Virginia as the seventeenth century dawned with thoughts of trade and hope of gold. The wave of colonization thus inspired in the early seventeenth century contained elements of the utopian concept as well. In this context, Hariot's words and More's vision shortly were to acquire a new significance.

2

Metals in the Early Colonies

SAUGUS FURNACE 1652

You brave heroic minds worthy your country's names,
That honor still pursue; Go and subdue!

Success you still entice to get the pearl and gold,
And ours to hold
Virginia, Earth's only paradise.—Michael Drayton, 1606[1]

Turbulent events in Europe at the end of the sixteenth century once more occupied England's men and resources, producing another lull in English attention to the New World. In the aftermath of the Armada, British sea power steadily increased, a necessary factor if colonization were to succeed, as the consequences of the failure to supply the Roanoke colony had demonstrated. A new century brought changes in the prevailing conditions. The Stuart monarch, James I, began his reign in 1604 by negotiating a peace treaty withdrawing English forces from the Continent. Growing in power and confidence, the English were once more free to turn their thoughts westward. The old motives for colonization still had their attractions—to search for the yet undiscovered but suspected treasures of gold or silver, to serve as jumping-off points to locate the equally elusive Northwest Passage to China, or to provide bases for harassing the Spanish in the Caribbean. But in the early 1600s there was added a growing conception of foreign colonies as commercial enterprises to trade with the natives and to exploit natural resources required to feed the growing prosperity of England. The settlement of Jamestown and its early history contain elements reflecting all of these themes. Lacking, however, was a clearly defined sense of permanent settlement, as envisioned by Gilbert. Norumbega, not utopia, was the primary inspiration drawing Englishmen to Jamestown.

The Jamestown colonists found neither gold nor a sea route to the Indies. The old antagonisms toward Spain cooled, although for another century the Spanish regarded warily the southward spread of English settlement toward their own Florida colony. When the anticipated trade also failed to develop, the colony nearly collapsed. Yet the colonists stayed, and other Englishmen followed them to swell the population and spread the area of permanent settlement. Two decades after Jamestown, another stream of settlers arrived in Massachusetts Bay, fleeing the religious and civil strife fostered by the intransigent policies of the Stuart monarchs. Whereas Jamestown had been founded to exploit the resources of America, the new colony was conceived in utopian terms, as a New Jerusalem of perfect faith and harmony in the wilderness. With the settlers south and north came English civilization, and with the concomitant practices and artifacts of that civilization came the need for iron.

Sir Thomas More had, in one sense, already defined the path the European nations would have to follow to establish and maintain permanent settlements in the New World. Utopia had lacked the metals essential to Western civilization; therefore they had had to be imported. If the metals necessary to support European material culture were not to be found in the colonies, Europe would have to supply them, and, though gold and silver were desirable, iron was indispensable.

It should not be surprising, therefore, that within a few decades, both Virginia and Massachusetts were the sites of ambitious ironworks, but each venture reflected the corresponding values present in the founding of the colony. The Virginia ironworks at Falling Creek were started in the spirit of mercantile exploitation, to provide iron for England. The Company of Undertakers of the Iron Works in New England, which financed the Hammersmith works in Massachusetts, was based on the entrepreneurial inspiration of a colonist, John Winthrop, Jr. Although basically English- and profit-oriented, the company had the active support and encouragement of the colony and its leaders, who perceived the need for an American ironworks to serve the permanent community they had founded. The Massachusetts experience pointed in the direction the history of metals in America would take henceforth.

The letters-patent to Sir Thomas Gates and others granting license to "make habitation, plantation, and to deduce a colony of sundry of his people into that part of America commonly called Virginia" were signed by James I on April 10, 1606. In accordance with a custom dating back to the reign of Henry VII, article IX of the patent granted permission to search for and mine gold, silver, and copper, reserving to the crown a fifth part of the gold and silver and a fifteenth part of the copper. Articles XI–XIV discussed the details of the provisioning of the colony; article IX contained the only explicit reference to potential products from the colony—precious metals.[2]

With the receipt of its patent, the Virginia Company of London immediately began to promote and organize an expedition to establish a colony. Although the same patent also authorized the creation of a

Virginia Company of Plymouth, with a land grant farther to the north, the Plymouth Company was much slower to act. By late fall 1606, the first group of colonists was ready to leave for Virginia. Early in December, His Majesty's Council for Virginia issued a set of orders to direct the actions of the expedition, including instructions that after landing the party was to be divided into three equal groups—one to build and fortify a village, a second to begin clearing and planting land for the support of the colony, and a third to explore the surrounding country. Specifically, the third group was to note any high lands or hills and to carry "a half dozen pickaxes to try if they can find any minerals."[3] The company also was encouraged to note any streams that might flow westward to the East India Sea; the search for a Northwest Passage still was very much intertwined with the search for gold; either course held the promise of easy profit.[4]

Certainly the Jamestown colony was not founded for the sole purpose of discovering gold or silver in America—that hope was only one of several that inspired the merchant adventurers who comprised the Virginia Company of London. For this study with its focus on metals to the near exclusion of other factors, however, the discussion of the search for gold assumes added significance. It represents the first stage of a pattern which the development of metals technologies followed in America, and the key to that first stage already has been noted—the idea of exploitation. It was a continuation of the same mind set that dominated the English voyages of exploration in the sixteenth century. There was no thought of developing the resources of America for people establishing permanent residence in America. The principal goal of exploration and of the initial efforts at colonization was to realize profit as soon as possible to enhance the power and prestige of individuals and thrones in Europe. Gold and silver simply represented the most immediate means to achieve that end.[5]

On December 19, 1606, the company sailed for Virginia. True to their charge, after deciding on the site of Jamestown for their settlement, Captains Christopher Newport and John Smith with twenty men went off in search of the headwaters of the river on the shores of which they had landed. Traveling inland to within sight of the Shenandoah Mountains, the party failed to discover any westward-flowing rivers. But, as Smith later recorded, they found that the waters of the streams running downward from the mountains to the west "wash from the rocks such glistening tinctures, that the ground in some places seemeth as guilded, where both the rocks and the earth are so splendent to behold *that better judgements then ours might have beene persuaded, they contained more then probabilities.*"[6] Under the influence of just such persuasion, Newport, upon returning to England in late July 1607, shortly after landing at Plymouth, wrote to Robert Cecil, earl of Salisbury, that Virginia "is excellent and very rich in gold and copper." He had brought a sample of the gold with him, Newport went on, and "I will not deliver the expectance and assurance we have of great wealth, but will leave it to your Lordship's censure when you see the probabilities."[7] But once more an "ore" sample proved to be worthless, and the error made by Smith and Newport soon became widely known. (Salisbury's censure is nowhere recorded.) A few weeks later, Dudley Carleton wrote to a Mr. Chamberlain reporting that Newport had recently returned from Virginia, but, he noted, "silver and gold have they none."[8]

The search was not ended by this disappointing development. Newport returned to Virginia bringing with him additional colonists, including two goldsmiths and two refiners. Their presence created a stir among the company, many of whom abandoned their regular duties. "There was no talke, no hope, nor worke, but dig gold, refine gold, load gold."[9] Yet, again, quantities of supposed ore were gathered, and it was proposed to fill a ship with it for transmission to England. John Smith, however, concerned over the direction events were taking, became determined to establish the colony on a self-sufficient basis before the gold mania destroyed the venture. Whether his prudence was further conditioned by a memory of the Frobisher episode is not known, but at Smith's urging the ship finally returned to England laden with marketable cedar wood.[10]

In the account of his adventures written in 1612, Smith expressed the reason for his reservations that immediate wealth was not certain. "There wanted good Refiners," he said, but noted, "Only this is certaine, that many regions lying in the same latitude, afford mines very rich of diverse natures. The crust also of these rockes would easily persuade a man to beleeve that there are other mines than yron and steele, if there were but meanes and men of experience that knew the mine from space."[11] Once again, the inability of the English correctly to identify valuable ore was demonstrated. The price of failure at Jamestown, however, was higher than that on the earlier expeditions. False rumors of successful strikes diverted the attention of the colonists from seeking dependable means of existence and threatened to bring about the dissolution of the colony.

Their weakness in prospecting notwithstanding, rumors of gold and silver kept the colony in a state of constant turmoil. One William Faldoe, for example, claimed to have found a silver mine not far from the Jamestown settlement, and his story created a stir when related in England. Again, help was dispatched to exploit the find, but poor Faldoe died before he could point out the site.[12] Smith's efforts to encourage agriculture were unproductive. No commercial commodities not already available at a far lower price in England or from nearby Europe issued from the colony, relations with the natives were very poor, and, still totally ignorant as they were of the vast stretch of land between Virginia and the Western Ocean, the sea route to India had not been found. By 1610, the various hopes that had inspired the foundation of the colony all were unrealized and enthusiasm in England for its continued support was at a low ebb. In Virginia, consideration was given to abandoning the venture. Sir Thomas Gates was preparing to evacuate Jamestown when Lord De La Warr arrived with a relief expedition, saving the colony from dissolution for the moment.[13]

In 1608, while the gold fever at Jamestown still raged unrestrained, a sequence of events had occurred that was to have increasing significance for America in the years to follow. On his second return to England, Newport had included as part of his cargo a quantity of iron ore. More ore was brought back on the third trip, in the fall of 1608. In the absence of sufficient quantities of marketable products, the ore probably helped to ballast the otherwise nearly empty ships. Early in 1609, the East India Company purchased seventeen tons of iron smelted from it at £4 a ton and found the quality satisfactory. The metallic wealth of America so long anticipated by Englishmen had first appeared in an unexpected form.[14]

From the founding of Jamestown, it had been necessary to supply the colonists with utensils, tools, and bar iron for a variety of uses. Mining was expected to be one form of industry at Jamestown, requiring iron picks and spades. Early on, an attempt was made to introduce the manufacture of glass, which also used a variety of specialized tools. Above all, there was the necessity to develop agriculture to sustain the colonists and even to provide some exportable commodity if mines and manufactures failed. A small but steady demand for iron tools and implements existed within the colony. Since the basic orientation of the colony was toward exploiting the resources to be found in America, the material needs of the colonists were supplied, piecemeal, by shipments from England during the early years. Tools alone were not sufficient, however. Blades had to be resharpened, broken tools repaired, any number of items manufactured. The colony needed skilled craftsmen.

As early as 1607, James Read, blacksmith, arrived in Jamestown.[15] Besides goldsmiths and refiners, other metalworkers followed. In 1610, for example, the Council for Virginia advertised for a number of skilled workmen needed in the colony, including "Iron-men for Furnasse and hammer," mineral-men, and gun-founders.[16] It is not known whether any metalworkers were among the relief company led by Lord De La Warr, but on his arrival he sent back word that, among other appointments designed to shore up the colony, Captain John Martin had been nominated master of the "workers for steele and iron."[17]

Small quantities of iron for use by the colony probably were reduced from ore very early in the colony's history. One dubious account exists of the production of iron in the first years. Francis Maguel, an Irishman in the employ of Spain, sent a letter to the Spanish Council in 1610 in which he reported having visited Virginia. He warned his employers that machinery already existed to process iron that could be used to make many ships that in the near future, could do much injury to the king of Spain. It appears, however, that Maguel was eager to report what he thought might be of interest rather than hard facts, for in the same letter he told of recovering an ore sample from the Virginia mines that yielded "three Reales of gold, of five in silver, and of four pounds in copper."[18] All this was at a time when, as Smith noted, the situation in the colony had become so desperate that it was almost abandoned. Far better concrete evidence has been provided by archaeological excavation of the Jamestown site, which revealed the possible presence of an operating forge by 1620. Some smelting of iron may have occurred before that using a kiln-pit, a method of reducing iron dating back to pre-Roman times.[19]

Still, the idea of producing iron in Virginia was slow to develop, although there was increasing reference to the abundant presence of iron in the literature concerning the colony after 1609. One prominent account by William Strachey, *Historie of Travell into Virginia Britannia* in 1613, mentioned the "excellent" mine from which Newport had obtained his iron ore. Strachey, however, like nearly all writers of the time, displayed excessive optimism and credulity concerning the mineral potential of the colony, favoring rumors of exotic metals over the substance of mundane ores like iron. In the same passage, noting the presence of iron at Jamestown, he gave full credence to the story of Faldoe's silver mine and to two other supposed silver mines close by. Also claimed was the presence of a mine of antimony "which . . . never dwells single but holdes assured legue with Quicksilver." Strachey then related that the natives obtained copper from the hills to the northwest, parting it from the stone without fire "and beat yt into plates, the like whereof is hardly found in any other part of the world."[20] Such accounts were a continuation of the gold-seeking mentality of the sixteenth century, reflecting a sense of impermanence. More significantly, as long as such thinking pervaded attitudes toward the colony, considering only temporary exploitation, the need to transfer to America the technologies that provided the material base for civilization remained minimal. As late as 1618–19, metal tools and hardware such as hatchets, hammers, chisels, hinges, scales, and pots figured prominently in a list of supplies being sent from England to the Jamestown settlers.[21]

The disappointing returns from Virginia steadily dampened the enthusiasm of investors. Some furs obtained by trade with the natives and, following the example set by John Smith, shiploads of timber, cedar, and walnut provided poor compensation for heavy expenses needed to sustain the colony which by April 1618 numbered about four hundred persons. Abundant iron ore and charcoal to process it offered an attractive alternative to the increasing exports of tobacco when Sir Edwin Sandys took over the important post of treasurer of the Virginia Company in 1619. The manufacture of iron would relieve the company of the costly need to supply the colony with tools and utensils, for which it would be self-sufficient. The anticipated surplus production from ironworks would find a ready market in England, where the expanding use of iron clashed with the need to restrict the cutting of forests used for fuel.[22]

Before 1700, all iron was produced by reducing the ore with wood charcoal. The lack of sufficient quantities of wood to meet the growing demand for iron had forced curtailment of the industry in some localities in England as early as the fourteenth century.[23] In the sixteenth century, a second heavy demand had been imposed on English forests—the need for timber to build the naval and, later, the commercial fleets. The charcoal shortage had been partially alleviated by the practice of coppicing (planting thickets from which to make charcoal), but the presence in America of large quantities of good ore and the seemingly endless forests to supply charcoal, noted by Hariot as early as 1585, offered another way to lessen the strain on the domestic resources of wood. The "Nova Britannia" of 1609, a piece of propaganda filled with extravagant claims concerning "this earthly Paradise," offered as one of its more realistic observations that Virginia had "Iron and Copper also in great quantitie, about which the expense and waste of woode, as also for building Shippes, will be no hurt; . . . the great superfluity whereof, the continuall cutting downe, in manie hundred yeares, will not be overcome."[24] Another tract, "A true Declaration," written in 1610, considered the threat of the diminution of the English navy for lack of wood, in part because of the great demand for charcoal in iron manufacture: "When therefore our mines of Iron and excesse of building, have already turned our greatest woods into pasture and champion, within these few years; neither the scattered Forests of England, nor the diminished Groves of Ireland, will supply the defect of our Navy. When in Virginia there is nothing wanting, but onely men's labours, to furnish both Prince, State and merchant, without charge or difficulty."[25] The reorganized company had ample reason to propose the manufacture of iron in the colony.

In 1619, major efforts to produce iron were launched by two groups. The adventurers of Southampton Hundred, apart from the Sandys administration, in that year were presented with an anonymous bequest of £550 for the conversion of Indian children to Christianity. After some debate among themselves and with the company, they decided to increase the amount from their own pockets and to use the money to establish an ironworks, part of the profits of which were to be reserved for the original purpose of converting the Indians. A Captain Bluett was sent off with a company of eighty men, but the captain died shortly after reaching Virginia. His loss constituted a great setback, but the records noted that "since that time, care had been taken to restore that businesse with a fresh supplie." The subsequent fortunes of the Southampton venture are unknown.[26]

In the next two years, there was considerable activity on the part of the company. One hundred fifty or more workmen, all purportedly experienced, were sent to the colony to construct three ironworks. Whether these were to be three separate operations or, as would seem more likely, an integrated unit containing a blast furnace, a forge, and a chaffery is a point of some confusion in subsequent discussions.[27] There appear to be no surviving records to indicate what efforts actually were made to erect such works, or where, or what problems were encountered in pursuing these endeavors. Not until over £4,000 had been spent and the company happened upon John Berkeley, a gentleman of reputed long experience in the iron industry, to take charge of the works was any real progress made. The Southampton operation apparently had collapsed by then or had been integrated into the larger project. The company took sole charge of the ironworks, for which a site had been selected at Falling Creek in what is now Chesterfield County.[28] On his arrival with a party of workmen late in 1621, Berkeley informed the company that a more suitable location for an ironworks could not be found, possessing as it did wood, water, mines, and "great stones hardly seene else-where in Virginia, lying on the place, as though they had beene brought thither to advance the erection of those Workes." Berkeley was fully confident that "by Whitsuntide next [1622] the Company might relye upon good quantities of Iron made by him."[29]

But on March 22, 1622, a few weeks before Whitsuntide, the Falling Creek settlement was massacred by Indians as part of a major uprising. The buildings were destroyed and the workmen's tools thrown into the river. One of the victims was John Berkeley. The surviving records do not make clear whether the furnace had been completed and any iron smelted. Archaeological evidence suggests that an integrated works, containing a furnace, bloomery, and forges, had been in the process of construction. Slag and artifacts recovered from the site indicate that some cast iron may have been produced and even converted to wrought iron by early 1622.[30]

Shocked though the company was by the news of the massacre, they responded with a letter of encouragement on August 1, expressing a strong intention to "againe resume that business so many times unfortunately attempted, and yett so absolute necessarie as we shall have no quiett untill we see it pfected."[31] But the problems entailed in financing and supplying further efforts proved to be insurmountable; the very next year the attempts to resurrect the manufacture of iron in the colony were abandoned. The failure of the Falling Creek project touched off a barrage of charges and countercharges at home. The large sums expended to foster ironworks in Virginia yielded no return other than an atmosphere of bitterness among the stockholders. The Sandys administration was accused of incompetence, with the opposition claiming, among other things, that too much had been attempted too fast in Virginia—several works where just one or two might have been successful.[32] No amount of argument, however, could disguise the fact that all manufacture of iron in Virginia had ended or that hopes for reviving that manufacture in the near future had all but ended also.

Five years later, the Stuart monarch Charles I expressed the prevailing sense of frustration in a letter to the governor and council of Virginia: "Our care for the welfare of our Colony in Virginia and of our people who have transplanted themselves thither doth much trouble us when we call to mind how many years are now spent since it began to be first planted and how little account can be given of any solid or substantial commodities which in all this time have been raised there."³³ His encouragement to find products other than tobacco, such as iron and "searching and trying for rich mines," fell on deaf ears. Although a few abortive attempts to resume iron manufacture were made over the next twenty years, the economy of Tidewater Virginia soon became firmly established on tobacco, an agricultural commodity.

The manufacture of iron at Falling Creek was conceived in the atmosphere of exploitation that was inherent in the creation and subsequent conduct of the Virginia Company. The impetus originated in England; organizations, finances, and supplies came from England, and, although it was hoped that the immediate needs of the colony for iron also might be satisfied, the product was intended for an English market. In that sense, the Virginia iron operations lay at the end of the chain of events begun with the voyages of exploration in the sixteenth century. The search for treasure had been superseded by the mercantilist goal, which was to obtain an essential commodity from the colony. But when the Jamestown colony turned to growing tobacco, it was because of the benefits to be derived by the settlers in America, not the company in England. Although tobacco culture produced a temporary cessation in attempts to find and refine metals in Virginia (with adequate income from tobacco sales, metal goods could be imported and the high cost of refining facilities avoided), the change in emphasis from what England wanted to what the colonists perceived as being in their interests marked a turning point in the development of America. The new direction was more discernible in the events taking place in Massachusetts.

The lands of New England stretching north and west from Narragansett Bay were more familiar by reputation to Englishmen at the start of the seventeenth century than those of Tidewater Virginia. These had been the rumored locations of Norumbega and Saguenay, the mythical empires of the sixteenth century abounding in precious metals. As late as 1614, Captain John Smith sailed from Virginia to the present area of Maine, "there to take Whales and make tryalls of a Myne of Gold and Copper." The only significant result of his visit was a detailed account of the region, an account more cautious and objective than those of his sixteenth-century predecessors. Smith noted the abundance of fish in the offshore waters and of fur-bearing animals in the forests which were to figure far more prominently in its trade than gold or copper, but he made several observations regarding metals. Whether gold and silver might be found in economic quantities he could not say for certain, but

Metals in
Colonial
America

THE INCONVENIENCIES
THAT HAVE HAPPENED TO SOME PER-
SONS WHICH HAVE TRANSPORTED THEMSELVES
from *England* to *Virginia*, without provisions necessary to sustaine themselues, hath
greatly hindred the Progresse of that noble Plantation: For preuention of the like disorders
hereafter, that no man suffer, either through ignorance or misinformation; it is thought requisite to publish this short declaration: wherein is contained a particular of such necef-
saries, as either private families or single persons shall haue cause to furnish themselues with, for their better
support at their first landing in Virginia, whereby also greater numbers may receiue in part,
directions how to prouide themselues.

Imprinted at London by FELIX KYNGSTON. 1622.

Colonial Broadside, "Inconveniencies"
The need for manufactured items in the early colonies was reflected in letters
home and in numerous broadsides such as this one offering advice to prospec-
tive settlers. Note the long list of tools and metal household goods that are
recommended. (*Courtesy, The Smithsonian Institution*)

he had no reservations about iron. "I am no Alchymist, nor will promise more than I know: which is, Who will undertake the rectifying of an Iron forge, . . . in my opinion cannot lose." That undertaking was not far in the future.[34]

In the years after the founding of Jamestown several colonizing ventures were attempted in the New England region. A small trading settlement on the coast of present-day Maine foundered in 1608 after a brief existence. Others at Wessagusett and Cape Ann suffered the same fate. Not until 1620 did the Pilgrims succeed in establishing a permanent community in the Plymouth area, maintaining a precarious existence supported by domestic agriculture and a limited fur trade with England. After the Pilgrims, a company of merchants called the Dorchester Adventurers reestablished a colony on Cape Ann in 1623, but by 1627 the group had dissolved. A remnant of the settlers held on at a nearby place they called Salem. With the exception of the Pilgrims, the early New England merchant-colonizers shared a common goal with the settlers at Jamestown. They all hoped to realize profits from the raw materials of the New World. In New England, however, their sights were directed to a search for the fish and furs Smith and others had reported. Sobered perhaps by past experience in the northern regions, they gave metals little or no priority among their stated goals.[35] About the same time that the Dorchester colony foundered, a grant in the Massachusetts Bay area was made to a group of Puritan merchants adopting the name New England Company, who, expressing the same conservative aspirations as their forerunners, dispatched Captain John Endecott in 1628 to collect furs and any natural products deemed marketable. A number of factors conspired to make the venture much different from the ones that preceded it.[36]

The Company had set about planning for the settlement and welfare of a colony even prior to obtaining the patent granting them the full powers to colonize and trade in the area. Major attention initially was given to the transportation of necessary supplies to America instead of products the colony could send home. Since metals were an inseparable and essential part of civilization, provision had to be made to procure them in adequate quantities for the settlers. When Endecott and a hundred followers sailed for Salem on June 20, 1628, among the items they carried with them as ballast, according to the first entry in the company records, were: "Iron 1 tun, Steele 2 ffagotts, Nayles and Lead 1 ffodder."[37]

Meanwhile, in England the dictatorial actions of the new King Charles I and the increasingly harsh policies of his favorite, Archbishop William Laud, against any forms of dissent from the Anglican religion caused growing dissatisfaction among the Puritans. Religious concerns soon transformed the intentions of the New England Company, whose royal charter was reaffirmed in 1629 under the title of the Governor and Company of Massachusetts Bay in New England. By August 1629 it was resolved to remove both the government of the company and its patent to New England.

The religious orientation of the Massachusetts Bay Colony, John Winthrop's vision of the colony as a city set on a hill so that all the world could witness their devotion to God and judge of their success in fulfilling His commission, is too well known to require elaboration. The word Puritan has the connotations of sobriety, industry, and strong commitment to duty inherent in its peculiar religious doctrine. But these also are the characteristics of stability and permanence, quite different from the sense of temporality that marked the early years of the Jamestown colony. There was a utopian idealism in the founding of the Massachusetts Bay Colony, but let it not be forgotten that utopians, as portrayed by More, had a strong sense of the realities of existence, the human need for work (six hours a day for all members of society) as well as recreation, and the material requirements of civilization. That same sense of hardheaded reality prevailed in Massachusetts. The religious motivation and social organization may have represented a departure from what were deemed to be the materialism and corruption of the Europe left behind. The cultural base of the new community, however, was characteristically European. To achieve and maintain permanence in America the material needs, to the extent that they coincided with religious precepts, had to be sustained. Yet for this sustenance, tools and other metal artifacts were needed. Iron axes were needed for felling trees and shaping wood to build houses, iron hooks and lead sinkers for catching fish, steel knives for trapping, and kettles and pots for cooking. Without reservation, the new utopians subscribed to More's adage that man, even in harmony with God, could not live without iron.

Initially, the company devoted much effort to supplying the basic material needs. During its first year, shipments to the colony included a broad range of metal artifacts, such as arms, ordnance, a wide variety of tools, and even household utensils. Of note in the latter category is an entry listing items to be sent in the future that includes pewter bottles, brass ladles, and copper bottles "of ye f[ren]ch making," the latter being a reminder that copper still was a metal sorely deficient in English stores.[38] But the colonists were aware that tools and utensils could be obtained only from England. Whenever possible, they were urged to take such things with them when they sailed. The Reverend Francis Higginson, first minister of the Church of Salem, wrote home shortly after arriving in Massachusetts in 1629 that all manner of livestock were needed and that people coming to the colony should bring with them cloth, leather, tools, iron, steel, nails, furniture, glass, and anything else that might easily be obtained in England but likely would not be available in Massachusetts.[39]

Between 1630 and 1640, more than twenty thousand people, the majority of them Puritans, emigrated to the Massachusetts Bay Colony. Although an increasing population presented a larger market for manufactured goods, they continued to be supplied from England in the early years, because this was cheaper when balanced against the relatively high capital expenditures necessary to establish self-sufficient industry. Part of the Massachusetts Bay Company's efforts to create a stable colony capable of developing some degree of trade had been the recruitment of skilled workmen to be included among the early settlers. Carpenters, masons, joiners, wheelwrights, and tailors were on the ships to America in 1628 and 1629. A little later, blacksmiths joined the colony. One of the earliest was Edmund Bridges, who arrived in the Massachusetts Bay Colony in 1636 and by 1641 was established in Rowley (now Ipswich), where his occupation was listed as "black-smith."[40] Initially, the new settlers brought with them enough wealth for the colony to purchase most of the supplies, including metals, that could not be paid for by trade in the staple commodities of furs and wood. As long as that condition prevailed, there was little incentive to develop a local metals industry.

The company assumed the responsibility for the basic provisioning of the colony, but expected a certain amount of return trade with which to offset the expenses of the growing outflow of men and materials. Unlike Virginia, where the constant urgings to find and mine valuable metals produced few results, the Puritans limited their commercial aspirations to the more mundane natural produce of New England. The early correspondence expressed the hope that quantities of salted fish could be available for return shipment, or beaver, or, at the very least, wood. In March 1629, however, John Malbon, "having skyll in Iron works," and Thomas Graves, who had self-proclaimed talent in the discovery of iron, lead, and copper mines, were hired by the company and sailed for Massachusetts in the same fleet with Reverend Higginson. Among its limited mineral resources New England was known to possess numerous deposits of iron ore suitable for development, but whether Malbon or Graves ever noted the presence of the earthy, brown limonite, called bog iron, that was prevalent throughout the coastal regions, is not known. Moreover, the interest in iron at such an early date produced no immediate benefits in light of the economic factors then prevailing.[41]

By 1640, however, the great influx of colonists was over, as conditions in England approached the verge of civil war and the end of financial and religious tyranny under Charles I and Archbishop Laud. The curtailment of immigration also dried up the vitally important flow of money to the colony, and as cash reserves dwindled without replacement, the colonists' credit with England began to collapse. Further aggravating the deficit problem was the near destruction of the fur trade through intensive hunting and expanding land use to accommodate the growing population. The manufactured products needed in the colony no longer could be obtained in the quantities desired.[42]

The crisis resulting from the failure of normal trade with England led to attempts to compensate for the loss by the development of internal resources. As early as 1641, a growing concern over the financial plight of the colonists prompted the General Court of the colony to pass laws to increase existing domestic manufacture and resolutions to encourage new ventures. Among the latter was the promise of a twenty-one-year grant of exclusive rights to the discoverer of any mine. The emphasis was on promoting any commodity suitable for export, including mica or graphite (black lead), in addition to metals, with the products to be shipped, preferably, directly to England.[43] By 1641, when the resolution was issued, it appears that several promising iron ore deposits had been identified, including those soon to be opened at Saugus and Braintree.

Since iron was one of the most essential materials the people of Massachusetts were required to import, the development of local iron beds promised to provide a double economic benefit to the colony, first by reducing the cost of imports and second by deriving income from the export of an anticipated surplus production to England. Whereas the argument for the manufacture of iron was similar to that used by the Virginia Company to promote the ironworks at Falling Creek, the impetus for the Massachusetts venture arose within the colony itself. To achieve the desired benefits required the erection of refining facilities, the ore itself having very little value, and the colony was in no position to undertake such an expense. The stage was set for the appearance of a new figure in the history of metals in America, the entrepreneur. When a mission led by Thomas Weld and Hugh Peter set sail for England in the summer of 1641, ostensibly to promote the interests of New England and to achieve a softening of the demands of creditors, one of its members was John Winthrop, Jr., a key figure in the establishment of an iron industry in colonial America.[44]

John Winthrop, Jr., the son of the first governor of the Massachusetts Bay Colony, almost certainly had no firsthand knowledge of iron production before he sailed to England, but he was a man of unlimited energy and vision. The processing of iron in the seventeenth century, from ore to the finished product, required a high degree of skill and knowledge that could be obtained only by the expenditure of long hours at the furnace and forge. The extensive body of theoretical and instructional literature that today assists the transition from study to practice is a modern phenomenon. Experience alone could assure the success of such a complex venture as iron manufacture before the nineteenth century. Winthrop, however, was a voracious reader with eclectic tastes. His broad interests in science made him well known to many of England's eminent natural philosophers.[45] His outstanding library, a large part of which has been preserved, contained numerous works on the sciences, including alchemy. The major sixteenth-century writings on prospecting, assaying, and refining metals by Agricola, Biringuccio, Ercker, and Barba are not among the surviving volumes of his library, nor is it known if Winthrop was familiar with them. Nevertheless, Winthrop probably was able to increase his interest in and general knowledge of metals from correspondence with friends in England and from lesser published sources.[46] His interest, coupled with a winning personality and strong connections because of the status of his family, made Winthrop the ideal person to promote the formation of an ironworks in Massachusetts.

While Hugh Peter, William Hibbins, and the Reverend Thomas Weld went about fulfilling the varied functions of their mission, John Winthrop, Jr., was hard at work promoting a scheme for the creation of an iron industry in America in addition to his official duties. Few details of how this was accomplished are known, but when Winthrop returned to Massachusetts in September 1643, he left behind in England the Company of Undertakers of the Iron Works in New England, an entrepreneurial organization created to support the manufacture of iron in the Massachusetts Bay Colony. Winthrop also returned with a number of skilled workmen and the tools and materials with which to begin the operation.[47]

After a careful survey of available sites, Winthrop decided upon a location in Braintree near good ore and with ample timber for charcoal. In early 1644, the building of a furnace was begun, and Winthrop petitioned the General Court for monopoly rights.[48] These were effectively granted for a term of twenty-one years at the session of March 7, 1643–44, with the provision "that within two yeares they make sufficient iron for the use of the country."[49] The words of the General Court of the Massachusetts Bay Colony introduced a new element in the evolution of metal technologies in America. The new utopia needed iron to be able to continue to function. For the first time there was a clear expression of the desire to produce metal—iron—in the colonies, for use by the colonists. No similar interest had appeared in Jamestown, where the emphasis throughout had been on exploitation for European consumption. The monopoly petition was just the first of many in the years to follow as the undertakers sought increasing concessions, which were granted by the court under the twin prods of the economic desirability of having a local iron industry and the timeless need to protect an investment that grew with each new concession.[50]

The Braintree furnace was completed by December 1644, although costs had been far in excess of estimates, a situation vexing to the absentee undertakers, who constantly were being prevailed upon to send more money. The estimates probably had been unrealistically low. Edward Neal Hartley has shown that the erection of a furnace for even £1,300 would have been an exceptional feat considering the high labor costs and the need to import skilled workmen and materials.[51] Some distance from the furnace at Braintree, a forge was constructed that cost nearly as much as the furnace installation. When the furnace was abandoned in 1647 for lack of ore, the forge remained in operation to process ingot coming from a new blast furnace erected by the undertakers on the Saugus River at Lynn.[52] Even before the Braintree furnace was completed, however, the redoubtable Winthrop had petitioned for the right to erect an ironworks at Pequot (near the present site of New London, Connecticut). Growing irritation with Winthrop's management finally led to his removal in 1645, and the operations were entrusted to an Irish ironmaster, Richard Leader. Winthrop's entrepreneurial spirit was not diminished by his experience in Massachusetts; he later attempted the organization of an ironworks in the New Haven Colony.[53]

Leader inherited a partially completed facility from Winthrop, but he sought a more suitable location for a works where all the components could be located in proximity to each other. He found such a site at Saugus. There the Hammersmith Works were erected, consisting of a furnace, two fineries, a chaffery, and a water-powered forge hammer. Later, a rolling and slitting mill were added. The "integrated" plant at Hammersmith incorporated the most modern technology for the production of iron and was, at the time, one of very few facilities in the world so completely equipped.[54]

Works similar to Hammersmith were built, singularly or in combination, with increasing frequency in the eighteenth century as the iron industry developed. The basic technology, however, remained essentially unchanged in America through the period of the Revolution. The reduction of iron from its ore could be accomplished by one of two methods. Small quantities of iron could be produced by heating ore with charcoal on the hearth of a bloomery forge. The resultant pasty ball or bloom contained metallic iron in a matrix of glassy slag and dirt. The iron was fused by repeatedly hammering the ball at red heat to break away the slag. For centuries, this "direct" process had produced nearly all of the iron used by Western man. At the start of the sixteenth century, an "indirect" method of producing molten iron in quantity had been perfected on the Continent and was rapidly adopted in England. The heart of the indirect process was the blast furnace.

A blast furnace performs two major functions in the production of iron. First, using charcoal both as fuel and to provide a reducing atmosphere of carbon monoxide gas and using limestone or some other rock melting at a relatively low temperature as a flux to carry away dirt and other impurities in the ore, the furnace reduces the oxide ore to liquid, metallic iron. Most of this is used in the form of pig iron, so named from the casting process which allows long molds to be filled simultaneously from a trench (the sow) leading from the furnace. The second function is the conversion of the molten iron into a variety of objects that may be cast in molds directly from the furnace, such as pots, firebacks, salt pans, and stove plates. Such items were a regular part of the output of the Hammersmith furnace, for sale to the colonists. The major furnace product, however, was pig iron.

The blast furnace at Hammersmith was about twenty-six feet square at the base, rising to a height of twenty-one feet. Later colonial furnaces achieved a maximum height of thirty feet, beyond which the weight of ore would crush the charcoal, blocking the passage of air.[55] The outer walls were of granite or other local stone, but the refractory stone of the inner chamber had to be imported. The identification of domestic stone capable of withstanding the heat of the blast posed a problem for furnace masters throughout much of the colonial period. At its widest point, the bosh, the inner chamber, was six feet in diameter. The lower part of the furnace had two arches, one of which admitted the tuyere or pipe carrying the air blast from water-powered bellows. Through the other arch, slag was drawn off of the pool of molten metal over the top of an iron-sheathed stone dam. When sufficient iron had been melted, the furnace was tapped by removing a ceramic plug in the dam. The furnace was charged by dumping ore, charcoal, and flux into the top of the furnace stack. The production of one ton of iron required three tons of ore and up to four hundred bushels of charcoal.[56]

To complement the furnace, a refining process was necessary, using separate hearths for refining, the finery, and for finishing, the chaffery. Pig iron usually absorbs a high amount of carbon from the charcoal in the smelting process, upward of 4 percent by weight, so that the end product is fairly brittle. It can be reduced to a more ductile form at a refining forge by reheating in air that slowly burns out the excess carbon. At the finery hearth, a pig was slowly fed into a charcoal fire. The high-carbon iron melted and dripped down into a pool on the hearth, some carbon being burned away as it melted and passed through the air. The process was repeated a second time and a third, after which, because the melting point had been raised with the removal of

Saugus Ironworks Restoration, Forge Shop and Rolling Mill
This view of the restoration of America's first major ironworks shows the stacks for the two refinery hearths and a waterwheel for a trip hammer on the left. The building on the right housed the water-powered rolling and slitting mill. *(Photographer, Richard Freer. Courtesy, United States Department of the Interior, National Park Service)*

carbon, the solid mass was shaped into a rough square bloom under a water-driven hammer. Control of melting and stirring during the refining stage to remove the right amount of carbon required exceptional skill and was unquestionably the most sophisticated operation in the manufacture of iron.

After the bloom had been further heated and hammered at the finery to produce a dumbbell-shaped form called an anchony, it was processed to finished shape by heating and hammering at the nearly identical chaffery forge. Here the ends of the anchony were reduced to uniform size. If a narrow-headed forge hammer was used, rotating the bar ninety degrees between strokes resulted in an elongated square or rectangular bar. Using a broad-headed hammer and reducing the thickness by striking in the same place produced a broad, flat plate. The hammering process also squeezed out most of the slag left in the pig. A little slag remained in the form of long, thin stringers that helped to give the material, now designated wrought iron, its characteristic toughness. Both hammer heads and anvils, each weighing several hundred pounds, were imported from England.

Hammersmith also had a unique facility in the combined rolling and slitting mill. Although the details surrounding the construction and use of this facility are sketchy at best, it was one of the earliest built anywhere and one of a very few built in America during the colonial period. Here partially finished bar iron was heated and reduced by passing it between heavy rolls. One waterwheel drove both rolls and slitters, sets of opposed discs with hardened steel edges that could cut the rolled bar to a desired width. The resulting rod was a suitable starting material for the manufacture of nails, a commodity much in demand for building in the growing colony. Iron emerged from the forge at Hammersmith, then, primarily as bar iron from which the blacksmith could form tools and other hardware items.[57]

Although the Hammersmith furnace reportedly produced as much as seven tons of cast iron per week, high expenses and no appreciable profits plagued the venture. By 1648, Richard Leader was complaining to the younger Winthrop that he was being ill-used by the undertakers and expected to be replaced.[58] In 1650, Leader was dismissed and John Gifford assumed the management. Whereas Winthrop initially had been overly optimistic, Gifford proved to be incompetent. He tended to be extravagant with company funds and often appears to have confused the company accounts with his own.[59] The facilities in full operation should have been able to produce far more iron than Massachusetts could absorb in its financially debilitated state, but Gifford was not able to approach that level of production. It is estimated that less than 170 tons of iron were made during his tenure. Iron had to be imported to satisfy the needs of the colony, and expenses continued to outstrip income. The undertakers began refusing to honor bills on the Hammersmith account when the iron to pay for them failed to appear, or to issue additional funds until they had received some return on the already large investment. Without money, the company could not sustain credit or pay its growing debts. No longer properly maintained, the ironworks fell into disrepair and disuse. Gifford's poor management was only one more aspect of an impossible situation. The venture collapsed in a welter of litigation in 1652.[60]

Sporadic production continued at Hammersmith into the 1670s. By 1676, the largest metalworking operation yet attempted in the colonies was abandoned, but the basic problem Hammersmith had brought forward for the first time remained unresolved. John Winthrop had realized the desirability of having ironworks in New England to supply a growing colonial demand for iron. Initially, the settlers had lacked the technical knowledge and the finances to build and maintain such works, so both capital and skilled labor had had to be imported to construct and operate Hammersmith. When poor management brought about the failure of the company, the reservoir of technological know-how remained. In the vacuum left by its demise, the colony struggled to meet its needs for iron by granting patents to towns for ironworks and by encouraging the individual efforts of the skilled workmen imported by Winthrop.[61] What originally had been conceived by the General Court as a convenient means to assist the ailing finances of the Massachusetts Bay Colony now had become a necessity, but few Americans were willing to risk investment in such costly and still unproven ventures. English investors, however, like the undertakers, thought primarily in terms of an English, not an American, market. The different perceptions of the function of an American ironworks were not to be resolved in the seventeenth century. The awareness by the colonists of the importance of metals, however, continued to grow.

The record to this point has emphasized the attitudes of Europeans toward metals in America to which the creation of Hammersmith ironworks offers no exception. The voyages of exploration were stimulated in large part by anticipation of gold and silver after the pattern of Spanish successes. A factor in the founding of the Jamestown colony was the hope of finding precious metals, and, when none were found, an abortive attempt was made to produce iron for export to England. Although the need for an American iron industry first became recognized in New England, the works begun by Winthrop did not fulfill that need. Rather, like Falling Creek in Virginia before them, they were an extension of the English iron industry in the New World, as their subsequent history displayed. The entrepreneurship that converted Winthrop's vision to a reality was English and remained so throughout the life of the operation. As Hartley has noted, the undertakers acted in the spirit of economic imperialism: "They looked forward from the beginning to the day when New England iron works would provide iron for export to England."[62] Hammersmith itself was a failure, but its very existence had helped to focus attention on the importance of iron, generating an atmosphere in the colonies that would encourage further attempts and give them higher chances of survival. A rapidly growing population in the colonies was becoming more consciously aware that it needed metals. Iron warranted the most attention as the colonies grew, both because of domestic needs and of changing events in England. Little was yet known about the potential supply of the others. England was to remain the source of most colonial metals for another century. Only one other metal, copper, received any significant attention by explorers and colonists in the colonial period. Many of the same elements that characterized the development of an iron industry were present in the search for copper.

Metals in
Colonial
America

3

Copper in the Colonies

*Mr. Endecott hath found a copper mine in his own ground.
Mr. Leader hath tried it.*—Governor John Winthrop, 1648[1]

The search for copper in colonial America was an extension of the preoccupation with precious metals that characterized the northern voyages of exploration in the sixteenth century and figured prominently in the early years of the Jamestown settlement. The elusive triad of gold, silver, and copper held continued fascination for the colonists even as their attention turned to the more pressing problems of ensuring an adequate supply of iron. That fascination was not limited to the English, as Cartier's voyages had shown. In the colonial period, it also was manifest in the actions of the Dutch at New Amsterdam and in the travels of French explorers in Canada and the west. Reports of silver and gold continued to filter from the backcountry of the colonies throughout the seventeenth and eighteenth centuries. The interest always generated by such rumors (they contained no real substance before 1799) reflected the tenacity of gold fever in America. The only ores actually discovered and developed to any extent during the colonial period other than iron were lead and copper. The interest in lead fluctuated because its primary uses, in fishing sinkers, musket shot, and window sashes, were easily and cheaply supplied by imports from England. Some of the attention devoted to lead was based on the experience in England, where lead ores were mined for their accompanying silver content. The small lead deposits in eastern North America could not be worked profitably for silver alone, however.

The search for copper in America was pursued by the Dutch and the French as well as the English. Workable deposits of copper were developed in several colonies, but no large quantities were produced. The consequences of the search for copper were more significant than simply the mining of the ore. The discovery of copper deposits in the Lake Superior region by French missionaries and trappers paved the way for exploration of the Mississippi watershed and French claims to the vast western lands. In the English colonies, it inspired some of the earliest attempts to explore the trans-Appalachian region in the Middle Atlantic colonies. Deposits of copper ore in Connecticut, New Jersey, and Virginia inspired entrepreneurial interest among important political and social leaders. For a period of time, the development of copper mining paralleled the manufacture of iron, although the production of copper never became a viable industry during the colonial period. Finally, the first steam engine in America was imported to drain water from the Schuyler Copper Mine in New Jersey, an event of greater symbolic importance for the development of the colonies than the production of copper itself.

The land between the Hudson and Delaware rivers was settled by the Dutch, and, until 1664, when the English under the duke of York seized the territory for a proprietary colony, this region was a foreign wedge between English settlements in New England to the north and around Chesapeake Bay to the south. In one sense, however, there was nothing foreign about the Dutch expectations of what they might realize in the New World. Their efforts toward finding metals and precious stones closely paralleled those of their English neighbors.

Although in both Virginia and New England interest in finding copper continued throughout the seventeenth century, the earliest attempts to mine it in America occurred in the land between these colonies while it was still under Dutch control. One of the twelve "Proposed Articles for the Colonization and Trade of New Netherland" drawn up for the Dutch West India Company in 1638 was devoted to the procedure governing the discovery of metals and minerals and the distribution of profits. The company was to receive one-fifth of the net profits from any minerals or jewels found on lands belonging to the colonists. Any discovery on unassigned lands would be the sole property of the company. The provisions were repeated in the statement of freedoms and exemptions to be granted "to all Patroons, Masters, or Private persons who will plant any Colonies or introduce cattle in New Netherland" set forth in 1640.[2] Within a few years of settlement, in October 1644, Henrick van der Capellen reported to the company deputies that minerals of copper, iron, and lead had been discovered and were sufficiently promising to warrant the sending of samples back to the Netherlands for testing.[3] The following year, additional specimens received from the Indians touched off a search to locate the site of the mines, reputed to be somewhere inland near the Raritan River, for the company's benefit. As had been true with the "gold" ores of the French and English before them, the Dutch ore samples proved to be worthless, but the interest in potential mineral deposits continued unshaken.[4] The pattern of events of the early years of the Jamestown colony was repeated by the Dutch in New Amsterdam. A search for gold or other valuable minerals in anticipation of rapid exploitation of the region's resources had been unproductive. That failure and the inability to develop any commercially satisfactory export produced a sense of discouragement, aggravated, in the Dutch case, by the closer control of the West India Company.

By 1649, the fortunes of the colony had begun to ebb. The colonists sent a petition to the home government in the Netherlands protesting the sorry state into which the commercial policies of the West India Company and bad fortune had driven them. The petition requested redress and support. Of the eight reasons cited by the colonists for their plight, one was the loss of the ship *Princess*, which contained "fully a hundred different samples of Minerals."[5] The *Princess* was the second Dutch ship that failed to reach home with mineral samples; a shipment sent out by Arent van Corness by way of New Haven and England had foundered earlier. The losses were deemed serious because, the colonists maintained, they were relatively certain that mines of quicksilver, copper, gold, silver, and iron had been discovered and tin and lead were present also. Only confirmation of the value of the finds was needed, together with more workers, to begin mining.[6] In severe economic straits, the colonists clung to the hope that efforts to mine any promising deposit, including iron, might produce badly needed revenue. Iron, however, held little interest for the West India Company. The anticipations of finding richer mines, as was also true for their English neighbors, continued to prove illusory.

Until 1659, no firm evidence was forthcoming that copper was to be found somewhere in the colony. That year, Claes de Ruyter reported that a copper mine existed somewhere between the Manhattans and the South River, gold also, and probably quicksilver. De Ruyter was dispatched to assist in locating the exact sites.[7] Subsequent events are obscured by gaps in the historical record. It is certain only that at some point after 1659 and before the British takeover in 1664 a road was constructed from the Dutch village of Esopus (now Kingston, New York) over one hundred miles to a point on the Delaware River just above the Water Gap, now the village of Pahaquarry. Two small copper mines were located there, but any attempts to work them had ended by 1664.[8]

The concerted search for copper occurred earliest in the New Jersey region, but Virginia and the middle and southern colonies also witnessed similar though less extensive activities during the colonial period. The sweeping arc of the Appalachians, from Pennsylvania to northern Georgia, long was regarded as the potential source for mineral wealth. When exploration pushed outward into the mountains from the Tidewater settlements, where the gold fever of the early Jamestown years still smoldered, there was the renewed anticipation of finding richer fare in the uplands.

In January 1642–43, the Virginia Assembly passed an act to encourage exploration of the region southwest of the Appomattox River toward the mountains with an optimistic provision reserving to the crown the fifth part of royal mines (that is, gold and silver).[9] Further orders were issued by the assembly in 1652, 1653, 1658, and 1659–60, promising fourteen-year monopolies to explorers who would "endeavor the finding out of any Commodities that might probably tend to the benefitt of the Country." Although the later acts do not specifically mention mines or metals, one may presume that these were among the implied beneficial commodities. Clarence W. Alvord and Lee Bidgood, in examining the economic motivation for the exploration of the trans-Allegheny region, have concluded that a search for mines was second in importance to the chief goal of establishing a fur trade with the Indians.[10]

In response to the acts of the assembly, several minor expeditions were made but found nothing of value. One of the first major explorations of the region was led by John Lederer into the mountains of western Carolina in 1669–70. In his report, published in London in 1672, Lederer noted the presence of copper ornaments among the natives. He was confident that further search would produce the discovery of rich mines to the southwest, and he voiced certainty that continued exploration in that direction soon would lead to the Spanish silver mines. Here was an echo of the words of Governor William Berkeley, who, describing the preparations for an expedition across the mountains in 1669 "to find out the East India Sea," had expressed similar hopes for finding silver mines on the way, "for certain it is that the Spaniard in the same degrees of latitude has found many."[11] The optimism of Berkeley and Lederer, spurred in part by an erroneous conception of the geography of North America, once again proved unfounded. Virginia's interest in metals slowly declined throughout the remainder of the seventeenth century as continued explorations failed to find gold. More easily detected by the green and blue coloration of its salts, some copper was discovered in the backcountry from Pennsylvania to Virginia, too far removed from the roads and rivers of the coast to encourage development. That failure represents perhaps the most ironic development in the search for metals during the colonial period. The eastern flank of the mountains in North Carolina, southwestern Virginia, and northeastern Georgia contained gold in some abundance. The richness of the rocks and streams was unrecognized until the beginning of the nineteenth century, when the area traversed by the trans-Allegheny expeditions became the focus of the first major gold rush in the United States during the 1820s.[12]

While the English colonists prospected up and down the eastern flank of the Appalachians for copper and other metals, one of the world's greatest deposits of the red metal lay to the northwest in the region around Lake Superior. Well known to French explorers, trappers, and missionaries who visited or passed through the region, it failed to be developed throughout the colonial period and well into the nineteenth century.

Centuries before the white man saw the coasts of the New World, the aborigines of North America carried on extensive mining operations on the Keweenaw Peninsula along the south shore of Lake Superior and on Isle Royale. Using primitive stone tools, they were able to dig for and raise native copper boulders weighing several tons.[13] Hammered copper artifacts from the Lake Superior region have been discovered in Indian graves as far away as Alabama and Florida,[14] although there is no evidence that the copper was ever heated to shape them. Many of the copper trinkets noted by the early explorers in Virginia and New England probably had arrived there by way of trade from the Superior region, and the Indians' awareness of the metal lodes "somewhere to the west" undoubtedly helped to create the legends of places such as Saguenay that tantalized Europeans in the sixteenth century.[15]

The first description of the Lake Superior copper region by a white man was the report of Father Gabriel Sagard, a Recollect missionary, who published in Paris in 1632 a journal of his visit to the Indians living about Lake Huron. After presenting a lengthy description of the customs of the natives and their need for spiritual succor, Father Sagard complained that his exhortations for assistance to effect one hundred thousand conversions fell on deaf ears, his listeners preferring news of how to obtain one hundred thousand crowns for their own benefit. He replied:

> Here then, you who are without religion, here are the treasures and riches to which alone you look with so much concern. They consist chiefly in quantities of furs, from various species of terrestrial and amphibious animals. There are in addition copper mines not to be despised, from which profit might be derived if there were a population and workmen who were willing to work them conscientiously. This might happen if colonies had been established, for about eighty or a hundred leagues from the Hurons there is a mine of copper, an ingot from which was shown me by the interpreter on his return from a journey made in the district. It is believed that there is some also in the direction of the Saguenay, and even that gold, rubies and other wealth are found there.[16]

The Jesuits participated most widely in the exploration of the Superior region, and their reports contained frequent reference to mines and minerals in the north central territory and along the Mississippi and its tributaries. Father Claude Jean Allouez, S.J., gave the first eyewitness account of the Lake Superior copper by a white man in the *Relation* of 1666–67: "On the second of September . . . we entered Lake Superior. . . . One often finds at the bottom of the water pieces of pure copper, of ten and twenty livres' weight. I have several times seen such pieces in the Savages' hands; and, since they are superstitious, they keep them as so many divinities, or as presents which the gods dwelling beneath the water have given them, and on which their welfare is to depend."[17] The *Relation* of 1669–70 carried a far more extensive description of the Lake Superior copper region gathered from Indian reports, including an account of the great abundance of copper to be found on Isle Royale, one of the principal sites of prehistoric mining activity. The numerous stories of the presence of copper all around the lake, together with samples in hand weighing over one hundred livres, convinced the writer, Father Claude Dablon, S.J., that "it [is] worth while to undertake an exact investigation in these matters," and a voyage of exploration was planned by the fathers for the following summer.[18] The following year, it was noted that "somewhere there are parent mines which have not yet been discovered" that must be the source of the many large lumps of copper found everywhere, including a "great rock of copper, seven or eight hundred livres in weight, seen near the head of the Lake by all who pass."[19]

Samples of copper were sent back to Quebec and on to Paris by the Jesuit fathers. These incited the interest of the authorities, particularly that of the intendant to the king, Jean Baptiste Talon, who encouraged a rigorous policy of exploration in the hopes of finding commercial deposits of metals. Talon's short reign as intendant, 1665–72, was a most active one as he attempted to foster the mercantilist theories of the French minister of finance, Jean Baptiste Colbert, by seeking out new sources of wealth for the colonies to contribute to France.[20] In 1668, Talon recruited the young explorer, Louis Jolliet, to lead with the half-breed, Jean Peri, an expedition to the Lake Superior region to discover the source of the copper the Jesuit fathers had reported to be there in such abundance. Although Jolliet and Peri had to report their failure to find the "source" of the copper, their trip established the usefulness of travel by the lake route to the interior. Subsequently, their route was used by French trappers to extend the fur trade when the English sealed off the Hudson Bay watershed in 1670.[21] In that same year, Talon dispatched a second expedition led by Daumont de Saint-Lusson to search for the copper mines and to take formal possession of the whole interior in the name of the king. Among the company were Jolliet and Fathers Dablon and Allouez. On June 14, 1671, at Sault Ste. Marie, Saint-Lusson read a formal proclamation that claimed possession of the lakes, adjacent rivers, countries, and streams from there to the western ocean and from the northern ocean to the South Sea. The expedition did not find the copper mines, but it gained, in the name of France, half a continent.[22]

With Talon's recall to France in 1672, the concentrated efforts to locate the main sources of the Lake Superior copper lode ended. Further notice of the presence of copper, lead, tin, and other metals continued to appear in the *Relations* of the Jesuits from locations as far away as Illinois and the banks of the Missouri. In 1687, the governor of New France, Jaques Rene de Brissay, the marquis de Denonville, wrote to Jean Baptiste Colbert, now the marquis de Seignelay, that hopes still existed for finding the body of the mine "wherefrom great advantages would accrue."[23] As late as 1700, the source of the Lake Superior copper defied discovery. In that year, an expedition of twenty men under the voyageur Father Pierre Le Sueur left Quebec to search for copper mines in the Minnesota region. Le Sueur sent back to France by way of the Mississippi River and Louisiana four thousand pounds of green and blue earth from several sites he discovered there, but the material turned out to be a worthless silicate of iron.[24] Le Sueur's expedition was the last serious effort on the part of the French to discover copper deposits that could be worked to the welfare of France. Based on the coherent mercantilistic theories of Colbert, high among which was the premise that colonies should endeavor to supply raw materials needed in France, the attitude toward the search for ores had been purely exploitative. The remoteness of the Lake Superior region precluded any attempt to develop the copper found there; no mines were ever dug. In 1739, Father Luc Francis Nau at Saint Louis on the Mississippi wrote in a letter to France, "An inexhaustible mine of copper has been discovered on the shores of Lake Superior, 700 leagues from here; but the profits will never be very great, owing to the immense expense of transporting the copper."[25]

The mining of copper in the English colonies in America began in the early eighteenth century with activities a few years apart in Connecticut and New Jersey. On June 21, 1671, Colonel George Cartwright, one of the New England commissioners for the Council of Foreign Plantations, delivered to the council a deposition affirming that the New England country was "healthful, fruitful and provisions plentiful, had store of good horses and doubtless lead and copper mines."[26] Over thirty years passed, however, before that optimistic prediction was fulfilled with the discovery of copper ore in the town of Simsbury, Connecticut.

Just who discovered the mine, or even exactly when, is unknown. The first recorded indication of a mine occurs in the minutes of the town meeting for December 18, 1705, which noted that a report had been made concerning the finding of a "mine of either silver, copper or mineral" eastward of the town (in what is now the town of East Granby, Connecticut). Two men were appointed to investigate the matter and return with details at the next meeting.[27] The report must have been favorable, for the town proceeded to reserve to itself forever all rights to the use and disposal of any mines or minerals. The citizens of Simsbury next took the unprecedented step of circulating a petition of subscription to which the names of nearly all of the inhabitants of the town, including children, were ascribed. When ratified, the petition assured each subscriber of a proportion of the proceeds in return for an obligatory share of the expenses.[28]

Undertakers were easily found to carry forward the development of the mine. The town granted the undertakers full control to mine and refine copper for a period of eight years, minus any time lost due to war (presumably with the Indians). Moreover, the town insisted that the proprietors and subscribers provide, at their own expense, an adequate site on a stream to serve the copper works, a highway to be built linking stream and mine, land grants for the use of the operation, and "full and free use of the adjoining timber or wood" both for building and refining.[29] All in all, it was a remarkably comprehensive and generous package of privileges and inducements to the establishment of an industry, second only to the patents and concessions given the Hammersmith ironworks in the colonial period.

The undertakers, in turn, issued a proposal, accepted by the town in 1707, that a tenth part of the proceeds should be set aside and divided, two-thirds for the maintenance of a schoolmaster at Simsbury and one-third "to the use of the Collegiate School errected within this Collony" (Yale College).[30] The connection between education and the proceeds of the mines was no accident. The original five undertakers all were members of the clergy, and the Collegiate School had been founded for the purpose of fitting men "for Publick employment both in Church and Civil State."[31]

Copper Mine Drift, Simsbury, Connecticut
Underground mining was rare in the American colonies and confined to the search for the precious metals. This horizontal shaft in the Simsbury Copper Mine, dating from the mid-eighteenth century, is preserved as part of the Old New Gate Prison and Copper Mine. (*Courtesy, Connecticut Historical Commission*)

By 1709, the activity in Simsbury had attracted the attention of the General Court of the Connecticut Colony. Unlike the custom in Massachusetts Bay, no charter or patent was issued to the company of undertakers. The court reaffirmed the rights of the proprietors to run the operation, but appointed William Pitkin, John Haynes, and John Hooker as commissioners to referee the growing number of disputes connected with the mines, particularly those instigated by individuals who had not signed the original petition and now wanted to share in the rewards. The commissioners were reappointed for a second term the following year.[32]

Simsbury and other towns north of Hartford along the Connecticut River Valley had originally been settled by groups of people who had broken away from the Massachusetts Bay Colony and moved westward. Basically, they shared the same Puritan values, including the sense of permanence and community that had been a distinguishing characteristic of the New England settlements. By 1700, a sense of community and of identification with the New World had become widespread throughout the English colonies, but these still were strongest in New England. Although the operation of the Simsbury copper mine originally exhibited elements of exploitation similar to ventures later organized in other colonies for individual gain, the resemblance was superficial. The town attempted to operate the mine for the benefit of the entire community. Although the only significant market in 1710 for copper was in Europe, the town arranged for the immigration of a number of skilled workers with their families from the German state of Hanover, primarily to build and operate a small refinery.[33] Such actions were in accord with those of the General Court of Massachusetts in promoting the building of the Hammersmith ironworks and with the later colonial developments in metals that were more obviously directed to the needs of the colonial market.

The second important development in copper mining in the early eighteenth century occurred in New Jersey, where Dutch prospecting had been intense fifty years before. Although the English had taken possession of the New Amsterdam territory, it remained for a Dutchman to make the major discovery of copper in the region. In 1710, Arent Schuyler, born in 1662 into a wealthy family of traders in the Dutch village of Albany, purchased a large tract of land on the eastern bank of the Second River (Passaic River) in an area known as New Barbadoes Neck. In 1712 or 1713, copper was discovered there. Within the next few years a mine was opened on the site and the ore packed in casks for shipment abroad. By 1717, Schuyler was making sizable consignments to his agents in the Netherlands.[34]

Questions soon were forthcoming over the destination of ore being shipped from New Jersey. In April 1721, Francis Harrison, surveyor at New York, wrote to the secretary to the Lords Commissioners for Trade and Plantations that from Schuyler's mine in New Jersey "there is Shipt on Board the Snow-Unity Robert Leonard Master for Holland one hundred and ten Casks of Said Copper Ore which we have not as I can find any law at present to prevent."[35] The letter prompted immediate action by the board. Harrison's letter was read with the minutes on June 14, and on June 20 the board secretary, William Popple, wrote to William Lourdes, secretary to the Lords Commissioner of the Treasury, enclosing a copy of it. Noting the absence of any suitable law to restrain the shipment of copper to countries other than England, he recommended "that this Practice may be of such Consequence to His Majesty's Revenue, that it do's desire to be consider'd in Parliament in order to be prevented by some act to be pass'd for that purpose."[36] England's interest in copper had hardly diminished in the years since Henry VIII had dispatched John Rut to seek metals in America. With the concurrence of the Lords of the Treasury the bill (8 George I c.18, Sect. XXII) passed in the House of Lords by a margin of 36 to 19 on March 2, 1722, among other matters containing a clause "to subject Copper Ore, of the Production of the British Plantations, to such Regulations, as other enumerated Commodities of the like Production are subject." The provision to enumerate copper was continued throughout the colonial period.[37]

The Schuyler mine continued to ship to England the largest amount of copper ore from the colonies, which by the 1720s had several operating mines. The Schuyler notebooks were examined by Thomas F. Gordon before they were destroyed by fire, and he found that nearly fourteen hundred tons of ore had been shipped to Bristol by 1731.[38] Bristol merchants who had extensive copper interests attempted to establish an exclusive right to the Schuyler copper but were rebuffed.[39] The New Jersey legislature joined the side of the merchants by passing, in 1733, an impost of 40s. per ton on copper ore not shipped directly to England. A debate ensued that finally resulted in the repeal of the impost by direction of the Privy Council in 1736.[40]

It is important to note that from the beginning the copper ore mined in New Jersey was not intended to fill any needs extant in an American market. Those needs were fairly small until well into the century because unalloyed copper was consumed at a rate probably not exceeding ten tons per year.[41] In fact, as late as 1770, when some copper was being used for coinage and the manufacture of brass artifacts was developing in Philadelphia and Boston, the practice still was to ship American ore to England for refining and to purchase the small quantities of raw copper needed.[42] Copper was used as the primary constituent of brass and bronze, but neither zinc (used in brass) nor tin (used in bronze) had been discovered in America, so that it continued to be easier to import finished artifacts than the raw materials. There is no record to indicate that the refining of copper ore to meet even the limited American market was ever attempted by Schuyler. His sole concern was to raise the ore for shipment to Europe, thereby acquiring foreign capital and credit.

The significance of the Schuyler mine was out of all proportion to the copper it yielded. Its history provided abundant evidence that the British thinking toward American resources had not changed in the more than one hundred years since the founding of Jamestown. The hearsay concern expressed over a single shipment from the Schuyler mine was sufficient to galvanize the British government to legislate the addition of copper ore to the list of enumerated articles. The attempts by British merchants to obtain exclusive rights to the Schuyler ore were predicated on similar interests. Moreover, the Schuyler mine was clearly conceived and operated in the seventeenth-century sense of economic imperialism, but here the entrepreneurship to effect the exploitation of the mines was coming from the colonial side of the Atlantic. The early success of both the Simsbury and Schuyler mines stimulated a renewed search for minerals, primarily in the Middle Atlantic region, and engaged the attention and investments of many influential men in the colonies.

The Simsbury copper mine was the first and largest in the New England colonies. Other discoveries of the red metal soon followed. In 1712, an outcropping of copper minerals was located in the town of Wallingford, Connecticut. Following the Simsbury precedent, the General Court confirmed the rights of the proprietors and their heirs to develop the mine.[43] With rumor of the discovery of yet another copper deposit in the town of Woodbury, the General Court in 1718 passed a general act to "promote the Improvement of the Copper Mines within this Colony" that set rules for the organization of companies patterned after the Simsbury model.[44] The Wallingford mine initially was suspected to be more extensive than the Simsbury lode, and hopes were high that "richer" minerals would be found at the site or nearby. But sporadic efforts at Wallingford continued only until about 1740, when the mines were abandoned.[45] Apparently, the Woodbury site was never developed.

As the Simsbury operation grew, it attracted speculators from other colonies and from as far away as Holland and England. In 1714, one Thomas Francis "of the City of London" purchased a part interest and, the same year, resold the share to John Alford of Boston at a price of £130.[46] The most prominent speculator and an entrepreneur after the fashion of Schuyler was Jonathan Belcher, colonial governor of Massachusetts from 1730 to 1741 and of New Jersey from 1746 to his death in 1757. In 1714 he purchased a one-sixth share in the mines, now including an operating refinery, for the sum of £421, a substantial amount, indicating that the property was considered to be quite valuable.[47]

As early as 1651, Governor Endecott of Massachusetts had petitioned the General Court of the colony for a grant of land to mine copper near the present site of Boxford, but only on the eve of the Revolution were any extensive diggings attempted in the area. Between

1720 and 1722, at least two mine pits were dug in the town of Topsfield, but the ore was of poor quality, and the effort soon was abandoned.[48] The absence of active mines in Massachusetts may have been a factor influencing Belcher's investment in the Simsbury operations. As was the case with Arent Schuyler, Belcher's primary concern with the mine was the personal profit to be realized by the sale of the ore in England. In Belcher's correspondence there was frequent discussion of the operation of the mine and references to shipments of ore to England via Hartford and New York in quantities of ten or more tons, for which he received credits of about £10 per ton. The expenses of mining the ore and of transporting it nearly ten miles overland to the river for shipment to England proved increasingly burdensome, however, as Belcher constantly lamented.[49] The refining operation built by the Hanoverians at Simsbury had been short-lived. It was closed in 1725 after the refinery and its contents were attached for outstanding debts.[50]

In 1729, the Board of Trade and Plantations drafted a series of "Queries Relating to the Colony of Connecticut" and sent them to the governor. Query number 9 asked, "What mines are there?" The reply of the governor and assembly in the following year was revealing: "There are some copper mines found amongst us, which have not yet been very profitable to the undertakers. Iron ore hath been found in sundry places, and improved to good advantage."[51] In 1773, the Connecticut General Assembly appointed a commission "to view and explore the Copper Mines at Symsbury, their situation, nature and circumstances, and to examine and consider, whether they may be beneficially applied to the purpose of confining, securing and profitably employing such criminals and delinquents as may be committed to them by any future law or laws of this Colony."[52] The colony purchased the balance of the unexpired leases for a sum of £60 and commissioned the subterranean caverns and buildings of the property as Newgate Prison.[53] During the revolutionary war, it held both military and civilian prisoners. In 1790 it was designated as the state prison, a function it served until 1897, thereby affording, according to one historian, "much greater advantage to the State than all the copper dug out of it."[54]

In New Jersey, the opening of the Schuyler mine was followed by the working of other copper prospects during the colonial period. With one possible exception, the search for copper followed the exploitative-entrepreneurial model established by Schuyler. Ore was mined and shipped to England. Close by Schuyler's lands, copper ore was discovered on the north bank of the Second River on land belonging to John Dod. For the next forty years, the mine was operated by Dod and various partners until flooding forced it to close.[55] Farther south, copper outcroppings were found along the flanks of the Watchung Mountains before 1754. The Bridgewater mine and a number of other excavations were dug in an unprofitable attempt to develop the sites.[56] For the most part, copper mining in New Jersey paid small returns for the efforts expended before the Revolution, but a lively interest in such ventures persisted throughout the period. The newspapers of Philadelphia and New York listed frequent advertisements for the sale of shares in mines and prospects and identified land for sale near current operations, unmistakably implying the possibility of further discovery.[57]

Of the many copper mines operated or attempted in New Jersey, two are of more than passing interest. One was located at Rocky Hill, where the Griggstown or Franklin Mine was opened about 1737, and the other on the banks of the Raritan River near New Brunswick, where a mine on the farm of Philip French was discovered in 1748. The Rocky Hill mine was notable for the illustrious list of men who held partnerships in the venture at different times. Among them were John Stevens, the father of Colonel John Stevens of Hoboken of steamboat fame; Robert Hunter Morris, son of the colonial governor of New Jersey, long-term chief justice of the province, and briefly governor of Pennsylvania; and William Allen, prominent Philadelphia merchant and founder of Allentown, Pennsylvania.[58] As were the Schuylers and Governor Belcher, these were men of wealth and influence attracted to the search for metals by the hope of personal gain. They, too, were interested in the profits to be derived from the sale of the ore to British merchants.

The French mine, however, represents a tentative approach to a stage of development more advanced than the investment-for-profit Rocky Hill operation. The full history of the French mine is not known, but for a period during its operation the works included a smelting furnace to refine the ore. Although the lower melting point of copper and its ores makes it easier to reduce to the pure state than iron, copper refineries were rare in the colonial period.[59] Some authors have advanced the argument that copper refining in the colonies was prohibited by English law. Considering the shortage of wood for fuel in England, such a law would have been anomalistic. Rather, it seems more likely that, as noted previously, the extremely small colonial market for copper made such facilities uneconomical. The presence of a refinery at the French mine, however, may have betokened some interest in selling to that market.[60] Were that true, the French mine would have been an exception to the pervading policy of the exploitation of colonial mineral resources.

In general, the successful New Jersey copper mines were situated near the coastal settlements and had easy access to the rivers and harbors. Inexpensive shipment of ore by water was necessary if copper mining was to be economically practical, as experience had shown at the Simsbury mine. Elsewhere attempts to develop deposits of copper ore invariably were correlated with ease of access to waterborne transport. In neighboring Pennsylvania, Sir William Keith, the proprietary governor, was one of the first to attempt to mine copper in present-day York County in 1724.[61] Although a relatively remote location in other respects, its proximity to the Susquehanna River held out some promise of profit. The venture, however, proved to be short-lived. It soon was followed by a more extensive operation about fifteen miles east of the river called the Gap Mine. Shareholders in the mine included (from 1732 on) Governor Morris of New Jersey and Thomas Penn, as well as William Allen, some of the same men with shares in New Jersey copper mines.

The Gap Mine was plagued by groundwater seepage, a common problem in copper mining, where the excavations along the slanting ore beds to any depth soon cut across fault zones down which surface water percolated to the water table level. In New Jersey, the Schuyler mine had suffered the same fate in the early 1740s, when the depth of the shafts and the inability to remove the water collecting in them caused work to be halted. In 1749, Benjamin Franklin visited the mine but recorded that it was not in operation.[62] John Schuyler, the son of Arent, who had inherited the profitable property, sent to England to obtain a "fire-engine" of the type designed by Thomas Newcomen which was used to drain water from the mines of Cornwall. This steam engine was imported along with several sets of spare parts, there being no industrial establishment in the colonies then capable of producing them, and assembled at the mine under the direction of Josiah Hornblower of England, the son of the engine builder. The engine arrived in 1753, but it took two years, until 1755, to complete the housing and assembling at an expense in excess of £1,000.[63]

Hornblower, who, like Schuyler, sought entrepreneurial opportunity, capitalized on his unique knowledge of the operation of the steam engine, without which the mine was worthless, to obtain a lease in partnership to operate the mine. All went well until a fire in 1762 destroyed the building housing the steam engine. The works were rebuilt and continued to operate until July 1768, when a fire of "suspicious origin" again destroyed the steam engine installation. Shortly thereafter, Hornblower and his associates abandoned the mine as no longer profitable.[64]

At the Gap Mine, when water threatened to halt work in the diggings, Governor Morris in 1755 proposed to Thomas Penn that the mine could successfully be maintained with the installation of a "fire Engine," similar to the one installed at the Schuyler mine. Morris's assumption undoubtedly was correct, but the expenses involved discouraged such a solution and the operations were ended.[65] By midcentury the manufacture of iron in Pennsylvania had become widespread and its economic advantages so obvious as to divert attention and investment away from copper and other metals.

Maryland had somewhat the same experience as its neighbor to the north. Minerals of some value did exist in the little-explored western part of the state, but they remained largely undeveloped throughout the colonial period. Governor Samuel Ogle answered the queries of the Board of Trade in 1748 by noting that there were "great shews of copper in many places, but of the several attempts that have been made to discover veins of that metal, none has yet been made that quitted cost."[66] The primary obstacle to development was the absence of navigable rivers, or even roads, by which to move the ore from the mines in the west to the coast for shipment to England.

In Virginia the development of copper mining in the eighteenth century still carried with it vestiges of the earlier "gold fever" evidenced by flurries of interest whenever the rumor of a discovery of gold or silver or copper reoccurred. William Byrd of Westover was a prominent Virginian who caught that fever. On May 15, 1712, Governor Alexander Spotswood wrote to the Board of Trade requesting a clarification on the amount of the royal shares in mines because he had "great reason to believe there are Mines lately discovered here," and on May 24, Byrd noted the rumored discovery of a silver mine in his diary.[67] Moreover, Byrd, as a member of the surveying party that established the Virginia–North Carolina boundary line, laid claim to a large tract of land along the state line, which he named Eden. No small part of his interest in the Land of Eden were his hopes "in the riches that might lie underground, there being many goodly tokens of mines," possibly silver, lead, and copper.[68] In 1733, Byrd, on a journey to explore Eden further, noted that in an area near the James River "there are many appearances of copper in these parts, that the inhabitants seem to be all mine-mad, and neglect making of corn for their present necessities, in hopes of growing very rich thereafter." Along the way Byrd took the time to examine Colonel Drury Stith's copper mine, to make inquiries concerning Colonel Cornelius Cargill's mine, and to visit the site of a mine he himself hoped to operate on Blue Stone Creek. The Stith and Cargill mines were small operations in the Tidewater region of southern Virginia. Byrd's tract, much farther inland and less accessible by water, was not developed, nor, for all his hopes, were the goodly tokens of the land of Eden.[69]

Of a more practical bent than Byrd was Robert Carter, another of the great Virginia colonial landowners with interests in copper. In 1728, while Byrd still dreamed of riches yet uncovered, Carter had organized the Frying Pan Company to mine copper in a tract in northern Virginia. Although the company procured twenty-seven thousand acres and made elaborate plans for their development, including the importation of Cornish miners, the venture proved unprofitable. It was abandoned, leaving only the Ox Road, which is the present site of Fairfax Court House, and the name Truro for the parish founded in 1732 in the hopes of "a new Cornwall in Virginia."[70]

Among the English, the prospecting stage in the development of metals in America had pretty much run its course by the early 1700s. In Virginia, a few men such as William Byrd and Governor Spotswood maintained a mild interest in discovering new mines and, as late as the 1770s, a venture organized by Alexander Henry in the Lake Superior region to extract silver from copper ores showed that the prospecting fever had not vanished.

By the Treaty of Paris in 1763, the great French holdings proclaimed by Saint-Lusson passed to the English. One of the first Englishmen to visit the Lake Superior region after France relinquished her claims was the fur trader, Alexander Henry, who arrived at the mouth of the Ontonagon River on August 19, 1765. Henry later wrote, "I found this river chiefly remarkable for the abundance of virgin copper, which is on its banks and in its neighborhood, and of which the reputation is at present more generally spread, than it was at the time of this my first visit." Later, he was led by the Indians to view an enormous mass of native copper which he estimated to weigh "no less than five ton."[71]

In 1769, Henry and an English trader named Bostwich formed a partnership with Alexander Baxter, who had traveled from England to the frontier post at Michilimackinac on the report of the copper ores in the region to form a company of adventurers for the working of mines. A petition for a charter was submitted which included on the list of subscribers (according to Henry) His Royal Highness, the duke of Gloucester, Mr. Secretary Townshend, Sir Samuel Tutchet, Mr. Baxter, consul of the empress of Russia, and Sir William Johnson, superintendent of Indian affairs in the colonies.[72] Johnson's feelings seem to have been ambivalent. When a letter from Baxter belatedly reached him requesting a draft of £55 because he had been proposed as a member of the company, Johnson wrote to London declining his share: "It is not in my power to find Sufficient Leisure for the dutys thereof to attend to my present domestic Concerns as I ought to do, much less to embark in any additional engagements however inviting."[73]

Returning to the Ontonagon region in 1771, Henry and Baxter deposited a small band of miners who commenced excavations in the side of a hill. After tunneling forty feet during the winter, they abandoned the effort in the spring when the clay walls of the tunnel thawed and collapsed. Operations were transferred to the north shore of Lake Superior in 1773, but the following year the company abandoned that effort, though not for the lack of finding copper. Sounding a familiar theme, Henry explained, "It was never for the exportation of copper that the company was formed; but, always with a view to the silver which it was hoped the ores, whether of copper or lead, might in sufficient quantity contain. The copper-ores of Lake Superior can never be profitably sought for but for local consumption. The country must be cultivated and peopled, before they can deserve notice."[74]

The prospecting stage never completely vanished from the American consciousness, however. Perhaps what Daniel Boorstin has characterized as the "go-getter" attitude prevalent in nineteenth-century America was nothing more than an extension of the dream of easy prosperity that preoccupied very early colonists in another form.[75] Certainly, the impatience for success that has been a recognized quality of Americans from the colonial period to the present has many of the same distinguishing features that marked the search for precious metals and the reaction to the tales of discovery in the much earlier period.

As the only metal other than iron to elicit widespread interest in the colonial period, the history of copper mining recapitulates the stages of development in metals during that period. Originally it was regarded as the least of the triad of "precious" metals sought by the Dutch, English, and French. During the seventeenth century, the search for the elusive triad drew explorers farther and farther inland. In the middle and southern colonies that search inspired the first trans-Allegheny expeditions. To the north, the French penetrated the Great Lakes region as far as the headwaters of the Mississippi. Little ore was found, and that was copper, but the knowledge gained subsequently helped to open the interior to settlement.

In the eighteenth century, the search for copper throughout the colonies repeated on a more limited scale the sequence of events in New Jersey. Initial explorations for metallic ores, which failed to discover the gold of Carolina, did detect the presence of copper. Although efforts were hampered by the lack of suitable transportation facilities in the sprawling reaches of America, that metal attracted the attention of entrepreneurs such as Governor Jonathan Belcher of Massachusetts, Sir William Keith in Pennsylvania, and Robert Carter in Virginia, who attempted to mine the ore for their own economic advantage. The exploitative stage in the development of colonial metals was very pronounced in the mining of copper. American experience in the production of metals was so limited that Carter deemed it necessary to import miners from Cornwall to aid his operation, and the citizens of Simsbury, Connecticut, needed German workmen to erect a copper smelter. Such importation of workers was often repeated during the century and helped to accelerate the transfer of metal technologies to the colonies. At the French mine in New Jersey and at Simsbury, refineries were built and operated briefly, producing copper that could be sold in the colonies, although the colonial market for copper was too small to make such operations profitable.

Also during the eighteenth century, the growing importance of iron, to which we now turn, overshadowed interest in copper. By the time of the Revolution, nearly all copper mining in the colonies had ceased. In the 1790s, copper mining was revived in the northern states. Old sites in Connecticut and New Jersey were reopened and new ones discovered. The greatest deposit of the red metal, one that had drawn the French westward to claim a continent, lay untapped and relatively unnoticed on the western border of the new nation. Finally, the Schuyler operation resulted in the installation of the first steam engine in America, an important step in the industrialization of the colonies. Years later, Nicholas Roosevelt, an associate of Hornblower when the mines were reopened briefly in the 1790s, erected steam-engine shops on land purchased from the Schuylers at Belleville. There he built the engines for the Philadelphia waterworks and, in 1798, an engine for one of the earliest American steamboats, the *Polacca*.[76] Unanticipated by Schuyler, the importation of his steam engine was perhaps the most important contribution of colonial copper mining to the future development of the country.

Metals in
Colonial
America

 4

Colonial Iron: The Birth of an Industry

To Virginia, the first of the English settlements in America, belongs the honor of inaugurating within her limits as a colony that most important industry, iron manufacture.—R. A. Brock, 1885[1]

The Saugus Iron Works . . . helped to lay the foundations for the iron and steel industry in the United States.—National Park Service, 1975[2]

The search for copper and other precious metals was but a romantic counterpoint to the prosaic but far more significant developments in the field of metals taking place throughout the colonies. The desire for a domestic supply of iron had first become evident in the terms of the support given the Company of Undertakers in the erection of the Hammersmith works by the government of the Massachusetts Bay Colony. In particular, the colony's needs were to be met before surplus iron could be exported. Even while Hammersmith operated, and after its collapse, other governments in New England offered inducements to build furnaces or forges to aid the welfare of their communities. Although many of the ventures thus attempted were short-lived, the precedents established in the densely populated New England colonies in the seventeenth century were taken up by Middle Atlantic colonies in the eighteenth as the population of America grew.

Meanwhile, events in Europe were creating a growing shortage of iron in England. Investors once more looked with favor on the establishment of American ironworks to alleviate the British shortage. They were joined by colonial entrepreneurs, who, daunted in their quest for immediate fortune in the discovery of precious metals, turned to the profits iron could bring in the expanding markets abroad and at home. The Hammersmith experiment was repeated by several groups of investors, British and American, with the greatest focus in the Chesapeake Bay area. The building of ironworks proceeded in spurts, in response to market pressures and the availability of investment capital. Well before midcentury, however, the center of iron manufacture in America had shifted to eastern Pennsylvania, where it was to remain for the next century. More significantly, the issue of whether to produce iron for sale to a British or an American market was being decided in favor of the colonies. The chain of events begun at Hammersmith had led to the creation of an indigenous iron industry in America. Its existence and subsequent growth greatly assisted the material development of the colonies and became a source of increasing concern to Parliament.

When and where did the American iron industry originate? The first attempt to produce iron in the colonies occurred in Virginia about 1619, but the short-lived venture at Falling Creek hardly qualifies as the birth of an industry. Its collapse marked the effective end to the manufacture of iron in the Chesapeake area for nearly a century. Secure in the income from agricultural commodities, the Tidewater planters imported metals and other manufactured goods. When in the early eighteenth century furnaces were once more built for refining ore, a common motivation did reappear, the impetus again coming from individuals and companies attracted by the profits to be derived from exporting iron to England. The later industry had no other bond of continuity with the earlier effort, however, and the needs of the colonial market were largely ignored.

A stronger case can be made for establishing the birthplace of the American iron industry at the Hammersmith works on the Saugus River in Massachusetts. The initial inspiration for the operation came from John Winthrop, Jr., a colonist, and important support was derived from the monopoly patent and other concessions granted by the General Court of the Massachusetts Bay Colony. During its term of operation, the output of the works was directed to the domestic market. Moreover, although the capital and organization were supplied by Englishmen and the undertakers were granted full export rights once the local needs had been met, the quantity of iron exported from Hammersmith was insignificant. Even before the works were abandoned, many of the skilled workers brought to Massachusetts by the company had become active in building and operating forges and furnaces elsewhere in the colonies.

John Winthrop himself promoted the manufacture of iron in settlements beyond Massachusetts Bay. Winthrop's checkered career with the Company of Undertakers had done little to dampen his enthusiasm or reputation. After severing his ties with the company, he enthusiastically laid plans for an iron furnace at New London.[3] Although that proposal never materialized, he did enter into a partnership with some of the leading men of the New Haven Colony to erect a smelter there. In the town meeting of March 1654–55, Stephen Goodyear asked "if any knew of the existence of iron ore" since Winthrop had arrived and could judge if the quality were good enough to warrant the erection of a works. Deposits of bog ore were common throughout the area, and Winthrop's opinions found ready acceptance.

A company was formed the following year and began building a small furnace on a stream separating the towns of New Haven and Branford, both towns sharing the supply of wood and costs in the ratio of three-eighths for Branford and five-eighths for New Haven.[4] In keeping with the practices established in Massachusetts Bay, the colony passed legislation to encourage the support of such operations, decreeing that persons erecting ironworks were to be free of taxes and that any partnership for the purpose would have limited liability for the partners so that claims against them were not to hinder the works in any way.[5] By the early 1660s the plant was in operation, helped along by men who had acquired their experience at Hammersmith.

Here even more than in Massachusetts, the furnace was a colonial venture. The construction of the works was financed on the American side of the Atlantic, and although hope for profit largely rested in the export of iron to England, the colony fully expected to benefit from the availability of a local supply of iron. As had been the case with Winthrop's choice of Braintree for the site of the first Massachusetts furnace, however, the Branford location proved unfavorable. The nearby ore bodies were too small for sustained operation, and costs ran higher than anticipated. Finally, no amount of effort could enable the works to overcome the financial and technical difficulties that plagued them, and by 1680 the site was abandoned.[6]

As a sidelight to the project it is interesting to note that in 1665 complaints were lodged with the town council by farmers living above the furnace. They protested that the dam that had been constructed to provide water power for the operation was drowning their fields and fences, one of the first examples in America of an ecological problem created by the development of industry.[7]

While Winthrop labored at New Haven, the Hammersmith works began to founder on the shoals of rising costs and weak management. For over ten years, the Company of Undertakers of the Ironworks in New England held a monopoly for the production of iron within the Massachusetts Bay Colony. In 1657, however, desiring to share the advantages of a local source of iron, the western towns of Concord and Lancaster jointly petitioned the General Court for permission to erect ironworks within their boundaries. The court, in reviewing the petition, noted that the company had failed at times to supply iron or had exceeded the stipulated price of £20. Therefore, "taking into consideration the great necessity of a constant supply of iron to carry on the occasions of the country, and being credibly informed that the works in present are not like long to continue," the court canceled the monopoly and granted the request of the towns to erect separate works.[8] There is no clear record to indicate that Lancaster ever exploited the privilege, but a bloomery was established in Concord by 1660. Subsequently, the Boston merchants, Simon Lynde and Thomas Brattle, obtained entire possession of the Concord works, which operated at least until 1682.[9]

During the time the Hammersmith plant existed, it required a number of skilled workmen and a larger body of unskilled help in a variety of functions. One estimate established from careful examination of the company records is that the nucleus of the work force was about thirty-five men, roughly a third of whom had specialties in the casting and working of iron.[10] The Concord bloomery was the first of a number of ironworks inspired by Hammersmith. Many were built or run by the same English workmen whom Winthrop and the Company of Undertakers had brought to New England to work at Hammersmith or by their descendants, and a large part of the legacy of Saugus stems from their efforts.

Joseph Jenks, for example, came to Massachusetts as a blacksmith at the time of the founding of Hammersmith. A skilled ironworker and metal craftsman, he operated a small private forge and trip hammer during part of his term of employment, where he converted bar iron purchased from the company into a wide range of manufactured items.[11] Joseph's son, Joseph Jencks, Jr. (the son appears to have initiated the variant spelling of his last name), joined his father for a time at Hammersmith, learning the rudiments of iron manufacture. After the company faltered, he was associated for a period in the 1660s with the Concord bloomery in which other Hammersmith men later held shares. In 1671, the son erected a sawmill and forge, the first ironworks in the Rhode Island Colony, near Pawtucket Falls. His success drew other settlers to the area. The forge was destroyed by the Narragansett Indians in March 1675–76 during the Wampanoag War, but the town of Pawtucket, founded by Jencks on the site, grew to become a center of colonial iron manufacture before the Revolution.[12]

The Leonard family became the most noteworthy of the several ironworkers from Hammersmith who were instrumental in initiating a number of new facilities in Massachusetts and New Jersey, passing down their skills from father to son for many generations. In October 1652, a group of forty-four men and two women in the town of Taunton, Plymouth Colony, formed a company to encourage the erection of a bloomery on Two Mile River, and Henry and James Leonard were invited to come from Hammersmith to the colony to operate it. The construction of a dam across the river and the need to import machinery for the hammer from England delayed the opening until 1656. Dependence upon England for hammers and other materials for furnace or forge construction was a recurrent problem in the erection of ironworks throughout much of the colonial period. By 1683, James's son, Thomas, was operating the forge, named Raynham, which then consisted of two hearths. As an indication of the prosperity to be realized from iron manufacture, the forge at times paid dividends as high as £4. 8s. a year on a full share of stock that had cost only £20, and the bar iron sometimes doubled as currency in the specie-poor colony. The value of the works declined in the early eighteenth century, but they continued in one form or another until 1876.[13]

In New Jersey a furnace was erected in or near Shrewsbury, Monmouth County, by at least 1674, when its operation was being conducted by Henry Leonard, a member of the famous family of New England ironworkers. The presence of iron ore had been noted by the early Dutch settlers of the New York–New Jersey area, but New York did not have an ironworks until well into the eighteenth century. Tintern Abbey, as the Shrewsbury Furnace came to be known, passed into the hands of Colonel Lewis Morris, lately arrived from Barbados, and his partners in 1676. By 1681, it contained both a furnace and a forge that produced iron into the early eighteenth century.[14] James Leonard remained in Taunton and later established the nearby Whittenton Forge which, after 1737, utilized high-grade ore imported from New Jersey.[15] Thomas and Nathaniel Leonard, Henry's son, also contracted before 1700 to operate a forge in Rowley Village (now Boxford) which proved unsuccessful.[16] In 1768, a survey of the ironworks in the Massachusetts Bay Colony listed eight forges and one furnace as being owned and operated by descendants of these iron pioneers. It can be said with some justification that when John Winthrop, Jr., first began to promote a company for the manufacture of iron in the Massachusetts Bay Colony, he was initiating a chain of events that far outlasted the Braintree and Hammersmith operations. Partly on the basis of these subsequent events, the restored ironworks were incorporated into the National Park System as the Saugus Ironworks National Historic Site on April 5, 1968.

In discussing the accomplishment of the heirs of Hammersmith, however, one point needs to be stressed. Viewed in the context of its inception, the attempt to manufacture iron at Saugus River had been as much of a failure as the earlier effort at Falling Creek. Neither venture had fulfilled the high hopes with which it had been founded. The period of sustained production achieved at Saugus was too brief and insufficient in quantity to pay for the high cost of constructing and maintaining the full-scale manufactory, and Falling Creek was destroyed even as production was beginning. The ironworks of Joseph Jencks and the Leonard family, which proved to be successful, had two characteristics that set them apart from projects such as Saugus or Falling Creek. They were far smaller in scale, the bloomery or slightly larger hammer forge being little more elaborate or expensive to build and operate than the simple blacksmith's forge. The technology also was less sophisticated, relying principally on the skill of a very few men instead of the complicated integration of the specialized services of a large work force, as was the case at Hammersmith. On the one hand was the age-old practice of the craft, on the other, the establishment of an industry.

Arthur Cecil Bining has suggested that the early failures were in some respects inevitable. He has listed "Numerous Lawsuits, Indian attacks, prejudice and opposition because of the large amounts of wood consumed by the works, and a preference for English iron" as some of the many reasons the colonial iron industry was so slow to develop in the seventeenth century.[17] But, as Edward Neal Hartley has observed, to counterbalance the reasons for potential failure, "There was ore. There was a market. There were the technical skills and the technical facilities. There was active interest and active effort from entrepreneurs, from 'government,' and from communities at large."[18] The positive factors cited are misleading, however. A market for iron did exist, but it was, in fact, not one market but many. No one colony in the seventeenth century was large or wealthy enough to support a full-scale ironworks, and neither communication nor transportation between the colonies was sufficiently developed to weld the small, scattered colonial markets into one. In Massachusetts, most attempts to manufacture iron were short-lived. Before the end of the seventeenth century, the ironworks organized and operated by the men John Winthrop had brought to Hammersmith succumbed almost without exception to the same fate that befell their larger parent. By 1676, Edward Randolph could report the existence of only six forges operating in the New England region, practically the only ironworks in the colonies.[19]

The erection of ironworks in the religious-agricultural societies of seventeenth-century New England produced unexpected consequences. The desire for local iron manufacture was strong, but its introduction raised a host of problems for which the colonial governments found themselves unprepared, and these problems also helped to undermine the utopian character of the Puritan colonies. The works required the expenditure of large amounts of money. In the case of Hammersmith, it was supplied primarily by English investors, but elsewhere ironworks monopolized appreciable amounts of the limited investment capital in New England. Workmen possessing the requisite skills but lacking the dedication to the rigorous and undiluted religious principles of the earlier settlers had to be brought into the colony. Those workmen added to the burdens of the colonies' governments.[20] Civil disorders became commonplace—drunkenness, stealing, swearing, and violations of the Sabbath—before the less religious workers and their families could be assimilated under the stern regulation of the Puritan governments.

The works required the creation and continuous reconsideration of a body of law dealing with manufacturing. To encourage ironworks, special privileges and dispensations were granted to investors and workmen that disturbed the equilibrium of the social unity so carefully nurtured in the founding of the colonies. Finally, a heavy load of litigation was created that occupied the courts when, one after another, the operations declared bankruptcy. There were disruptive consequences on commerce and the domestic economy when the works ceased to operate because the works did provide part of the colony's supply of iron goods. To these must be added the social effects of unemployment.[21]

An atmosphere conducive to industry was engendered even in failure, however. Before the end of the colonial period the thriving iron industry that developed throughout the colonies included many successful ironworks in New England. The true legacy of Hammersmith was its contribution to the future success of iron manufacture by stimulating interest and by initiating the importation of technical skills. On those contributions rest any claims to its being the birthplace of the American iron industry.

The New England experience of Hammersmith and its aftermath, the attempts to build other furnaces and forges throughout the region, showed that in the North at least the old desire to profit from the mineral wealth of America persisted throughout the seventeenth century. But to the old desire was coupled a new concern to secure a dependable supply of badly needed iron for local use. In the broader geographical context of colonial history, however, the full significance of iron to the settlers in America, and to the investors in England who provided most of the capital to build and operate the first ironworks, did not manifest itself until the early eighteenth century. There followed a new wave of investment and entrepreneurship by Englishmen and influential colonists, closely attuned to changing political and economic developments in England. In the early eighteenth century, several major ironworks were built that warrant inclusion among the pioneers of an American iron industry.

The first decade of the eighteenth century witnessed the protracted struggle of Great Britain and its continental allies to check the threatened dominance of France known as the War of the Spanish Succession. We tend to think of the twentieth century as the era of the world wars, overlooking the fact that the European conflicts of the late seventeenth and eighteenth centuries were also fought on colonial battlefields ranging from India to America. The War of the Spanish Succession also pitted the English colonies against the French and their Indian allies in what was known locally as Queen Anne's War.

The demands of war in England created increasing pressure for manufactured iron in an industry that had been experiencing a steady decline in the production of pig iron for over half a century. Moreover, the heavy use of charcoal in the operation of blast furnaces and forges placed severe limitations on the building of new works to meet increased demands. By the early eighteenth century, the major production of pig iron in England had shifted from the southeast, the earlier center, to South Wales and the West Midlands. The number of forges converting pig to bar iron also was concentrated in those areas. The West Midland forges actually required more pig than the region could produce. Throughout England, the output of furnaces and forges (approximately eighteen thousand tons of pig per year for the former and fourteen thousand tons of bar per year for the latter in the first decade of the century) was inadequate to supply the overall demand for iron.[22] England found herself increasingly dependent on imports from abroad, particularly from Sweden, Spain, and Russia, to fill the deficit. By the start of the eighteenth century, two-thirds of the bar iron used in England already was being imported from Sweden, and total imports exceeded ten thousand tons per year.[23] On the other side of the Atlantic, an expanding colonial population presented additional demands.

Events during the period immediately following the Peace of Utrecht (1713) further complicated the problem of assuring an adequate supply of iron to both England and America. On the death of Queen Anne in 1714, her distant relative George Louis, the Protestant elector of Hanover, ascended the throne as George I of England. Throughout the late war, England had maintained a flourishing trade and peaceful relations with Sweden. Hanover, however, was a member of an alliance of northern states waging active war on Sweden along the shores of the Baltic. Swedish armies had attacked Hanover, capturing Bremen and Verden. England's new king was Sweden's sworn enemy. At the opening session of Parliament in February 1717, the king laid before Parliament intercepted diplomatic correspondence that established the existence of a plot to incite rebellion in the kingdom. Jacobin supporters of the Young Pretender were to invade Scotland, backed by a Swedish force, as a prelude to an attack on England. On the basis of his evidence, the king asked for and quickly received from Parliament "An Act to enable his Majesty effectually to prohibit or restrain Commerce with Sweden." All trade was cut off forthwith.[24] When, after the death of the Swedish monarch Charles XII in 1718, peace was restored between Sweden and Hanover and the embargo lifted, an adequate supply was not assured. The peace was not complete; Russia continued to contest Swedish hegemony in the Baltic. Into the early 1720s, Russian armies ranged along the coast of the Gulf of Bothnia, a center for iron production in Sweden, destroying ironworks and mines and jeopardizing the supply of iron to England and her colonies.[25]

The earlier scarcity of iron became acute. While trade with Sweden was suspended, Parliament received numerous petitions protesting the effects of the embargo on the English iron industry. Some urged the encouragement of new investment in ironworks in the colonies to replace Sweden as a source of supply for England. As we shall see in a later chapter, this issue provoked increasingly heated debate among vested interests in England. The immediate effect was that as the imbalance between total British demand and a less than adequate supply continued to grow, the establishment of an iron industry in America, possessed of abundant supplies of ore and wood, became increasingly attractive to investors on both sides of the Atlantic.

The first works to be constructed in the colonies by English investors in response to the changing conditions in Europe was a bloomery built at the head of Chesapeake Bay shortly before 1720 (slightly east of present-day Perryville) by the Principio Company. A smaller private forge may have existed near the same site, but the first iron exported to England from the colonies, three and one-half tons, shipped from the Chesapeake in 1718, probably was the product of the Principio works.[26] The company was a consortium of English merchants, ironmongers, and ironmasters and included a few of the colonial gentry, among whom were Augustine and Lawrence Washington, father and brother of George Washington.[27] Thomas Russell, Sr., of Birmingham and Stephen Onion, major stockholders in the company, were sent to America to do the prospecting and to arrange the purchase of land. Onion, as ironmaster, began the construction of works and the stockpiling of ore and charcoal. Nearly six thousand acres of woodland along the bay were purchased to ensure an adequate supply of charcoal. The pace of development proved too slow for the investors, and Onion was recalled and replaced by John England, another experienced ironmaster.[28] Before he returned to London in 1724, on the same ship that carried Benjamin Franklin on his first voyage abroad, Onion had obtained options on thousands more acres in the Baltimore area and, in cooperation with the Washingtons, had made plans to build a furnace in northern Virginia. Within a few years, the Principio operations under England included a forge and furnace at the original site, and a second forge was being built at nearby North East, Maryland. The Accokeek furnace in Virginia was completed by 1728.[29]

The main product of the Principio Company was pig iron destined for export, although some bar iron was manufactured at the North East forge for sale in the colonies. In August 1727, the company ledgers showed that forty tons of pig iron were produced at a cost of £4. 5s. 9d. per ton. The price of pig at the furnace then was £10 per ton, and bar iron was quoted at £35 per ton. In keeping with the original interest, however, the major part of the output from the company's expanding facilities was sold in England.[30]

Shortly after the Principio Company began full-scale production, a second Maryland group was organized, also to feed the English market. This was the Baltimore Company, formed in 1731 with a capitalization of £3,500 sterling. Unlike the Principio Company, the partners of the Baltimore Company all were from Maryland—Charles Carroll of Annapolis, his brother Daniel, Walter Dulancy, Dr. Charles Carroll, and Benjamin Tasker. Within a few years the group had acquired extensive land, built a furnace and forge, and were actively engaged in a search for British markets for their products. To an even greater extent than the Principio Company, the entrepreneurs of the Baltimore Ironworks directed their output to England. From 1735 to 1737, nearly six hundred tons of pig and bar iron were sold at Greenwich to an agent of the great Crowley Ironworks at a price of £6. 5s. per ton, half of which was paid in cash with the balance in fabricated ironware. Keach Johnson has estimated that the company was able to realize a profit of from £1. 14s. to £2. 4s. per ton on the sale of its iron from 1735 to 1755. At its height, the company operated several furnaces and forges with a work force that included 150 slaves.[31] The Principio Company had been the first American ironworks to employ slaves, a practice that spread in the Chesapeake and southern Pennsylvania regions throughout the century.[32]

The entrepreneurial spirit that fired the creation and operation of the Principio and Baltimore companies was fully characteristic of the exploitation stage in the development of American metals, as noted previously. Principio remained an English company throughout its existence. The primary motivation governing its management was the profit to be derived from the sale of its products in England. Unlike Hammersmith, which had foundered under poor management, both the Principio and Baltimore companies prospered and grew. Principio acquired the rights to important ore deposits at Whetstone Point near Baltimore. To save the transportation costs of moving the ore to Principio Furnace the company established works nearer the mines. Kingsbury Furnace, also in Baltimore County, was purchased from Dr. Charles Carroll in 1751, when the company reached its maximum growth. In 1751, Maryland and Virginia exported 2,950 tons of pig iron to Great Britain, the great bulk of which had been produced by the facilities of the Principio and Baltimore companies.[33] Few directors of the British-dominated Principio Company noted the shift in values that occurred in the colonies after midcentury, giving more and more emphasis to the erection of American ironworks to serve American needs, not British interests. Thomas Russell II, however, was an exception. In 1771, he moved to Maryland, becoming an American in spirit as well as residence. During the Revolution, the state of Maryland confiscated the facilities and assets of the company as British property. The only residual rights recognized were those of Russell and of "a certain Mr. Washington, a subject of the State of Virginia."[34]

The Principio Company was the first eighteenth-century enterprise organized to exploit the rising demand for iron in England. At nearly the same time the same motive inspired the construction of ironworks in Virginia, where no serious effort had been made to mine iron since the destruction of the Falling Creek ironworks in 1622. This project was organized by another entrepreneur, Colonel Alexander Spotswood, who arrived at Jamestown in the summer of 1710 to take up official duties as the lieutenant governor. Within a few years, he was actively soliciting support to revive the long dormant production of iron in the colony. Spotswood first tried to interest the colonial legislature in reopening the old mines on the James River, but no assistance was forthcoming from that quarter. The fault rested, Spotswood complained, with the Tidewater planters, who construed their interests so narrowly that they could not see the benefit of an ironworks on the "far-off" James.[35] Appeals to the Board of Trade in England also fell on deaf ears. Undaunted, Spotswood acquired the support of a group of investors, as with the Principio Company largely English, and undertook the task of building ironworks himself.[36]

Beginning with a blast furnace on the Rappahannock River above Fredericksburg in 1719, the group completed two additional furnaces, one blast and one air draft, in the next ten years. To build and operate the works, Spotswood, like Winthrop before him, had to import skilled workmen. In Spotswood's case, they were Germans, recruited through a Swiss intermediary, and their settlement on the river was given the name Germanna.[37] A second air furnace was being built in 1732 when William Byrd of Westover, whose interest in copper has been noted, paid a visit to the Spotswood works at Fredericksburg and Germanna.

Byrd's record of the visit gives us the best picture available of the operation of an iron furnace complex in the early eighteenth century. With an eye to starting his own furnace, Byrd noted details on everything from mining and making charcoal to the final casting of the iron. The ironworks formed the heart of a "plantation," vast acreage including farms to support the community that operated the furnace, but primarily given over to woodland to ensure an adequate supply of charcoal. Byrd also recorded Spotswood's boast that he had been the first in North America to build a blast furnace, thus showing the way for the northern colonies.[38] In the literal sense, of course, Spotswood was mistaken. Figuratively, the colonel was not so wide of the mark; his endeavors roughly paralleled those of the Principio Company and slightly preceded the broader movement that saw the building and spread of ironworks in a number of northern colonies, led by Pennsylvania. But Spotswood was oblivious to the full complexities of that broader movement. In the entrepreneurial tradition, Spotswood looked almost exclusively toward England to find a market for his products. The colonel told Byrd that the price of iron was such that even with freight charges one might figure a profit of £3 a ton in England, a highly optimistic figure. But the output of Principio's Accokeek Furnace and a part of Spotswood's production were sent to the forges at the head of Chesapeake Bay for conversion to bar iron, and some of it was subsequently used for manufactured goods. Before 1750, all Virginia iron was cast, and for some years much of it was shipped to England in the form of sow and pig, there being no forge to work iron in the colony. Also Spotswood's air furnace then being constructed at Massaponax would, he told Byrd, provide a variety of cast-iron commodities "as cheap and as good as ever came from England." The last point was the only concession Spotswood made to the growing demand for iron in the colonial market. That the colonies' needs were not his concern was clear in his observation regarding Pennsylvania—"they have so few ships to carry their iron to Great Britain that they must be content to make it only for their own use, and must be obliged to manufacture it when they have done."[39]

Alexander Spotswood died in 1740 without comprehending the changes that had taken place in the manufacture of iron in America. No lasting heritage stemmed from his ironworking efforts in Virginia. The success of the Principio Company in Maryland inspired the building of other works; by 1749 the colony possessed eight operating furnaces and nine forges, but Principio, too, belonged to the exploitative phase in the development of colonial metals. That phase was giving way even as Byrd and Spotswood conversed. In the forests of eastern Pennsylvania and elsewhere throughout the middle colonies a truly indigenous American iron industry was being born.

If, in the early years of Jamestown, the talk, the hope, and the work was to dig gold, refine gold, load gold, in the early eighteenth century the talk, hope, and work on both sides of the Atlantic was to dig iron, smelt iron, manufacture iron. The colonies' reasons for manufacturing iron were different from those in Great Britain.

The same shortages that beset England also affected her American colonies. Lacking the facilities to produce for themselves, the colonists, too, had sought foreign sources for bar iron and manufactured ironware. Between 1701 and 1710, 1,960 tons of foreign iron and more than 700 tons of steel had passed through English ports to America. As early as 1704, Parliament had abolished the drawback on the reexportation of foreign ironware to the colonies at the insistence of London manufacturers who complained of the effects of foreign competition. A few years later, Parliament passed legislation requiring that all bar iron imported by the colonies must first land at an English port and be subject to high British export duties. British ironmasters had protested that access to cheaper foreign bar stock had encouraged the manufacture of tools and hardware in the colonies, where British goods were at a disadvantage of as much as £15 per ton.[40] The actions of Parliament complicated the efforts of the colonies to obtain an adequate supply of iron for their own use and thus could not help but precipitate a reconsideration on the part of the colonists of the need for additional forges and even furnaces within the colonies.

The availability from non-English sources of cheap bar iron of good quality for processing into finished products may, in fact, have created a barrier to the building of ironworks in the colonies, especially when coupled with the high costs for facilities. The British actions completely changed that economic picture. The colonists now were confronted by a sudden sharp increase in the price of iron in all forms. At the time English ironmasters petitioned Parliament for relief, the shortage also was felt in the colonies. Up to the early eighteenth century, the few colonial forges had supplemented an adequate supply of imported iron. The embargo on Swedish iron meant that the growing colonial demands could no longer be met by importation. Moreover, Parliament's action in forcing a rise in the price of so basic a commodity as iron seemed to be a clear indication that, to the British government, the welfare of a few British ironmasters took precedence over that of the American colonists.[41]

In the climate of uncertainty, the colonists began to act to fill their own needs by producing their own iron. In 1716, Thomas Rutter had built a bloomery forge in the wooded Manatawny region, about forty miles northwest of Philadelphia. A few years later, coincident with the cutting off of all supplies of Swedish iron to England and the colonies, he headed a company of colonists which in the same region erected Colebrookdale Furnace, named after the well-known English furnace of Abraham Darby. This was the first blast furnace in the province. In 1718, with the embargo still in effect, Samuel Nutt, Sr., began operation of a bloomery in the forest lands of Coventry, shortly followed by refinery forges and a blast furnace that became the Coventry Iron Works. Other forges and furnaces followed.[42] Unlike Principio, or Governor Spotswood's Virginia furnaces, the Pennsylvania ironworks directed their products to the colonial market. Within two decades, a score of works were operating, being built, or proposed in the Central Atlantic states and New England for the same purpose, financed or operated by men from all levels of colonial society. A major industry was being born in America, with the blast furnace as its symbol.

The men who created the new industry were, for the most part, Englishmen. Thomas Rutter and Samuel Nutt, the pioneers in Pennsylvania ironworks, both had been born in England. Thomas Potts, who built Mount Pleasant Furnace on Perkiomen Creek in 1737, was of Welsh extraction and a member of a family who passed down the skills of ironworking from father to son into the nineteenth century.[43] In Maryland, as Michael Warren Robbins has noted, the ironmasters and skilled workers were recruited in England for the several colonial furnaces and forges built within that colony.[44] In Pennsylvania, where Philadelphia served as a point of entry to the New World, ironmasters were predominantly English, but a significant number had other national backgrounds. Jacob Duk, forge owner, was from Zurich; Gerhard Etter and Henry William Stiegel had been born in Germany. The keys to establishing an indigenous American iron industry were the presence in the colonies of the requisite technical skills and the availability of colonial capital to finance the building of ironworks. By the early eighteenth century, the knowledge of the working of iron was widespread throughout the colonies. Not all of the early ironmasters, however, directly possessed that knowledge. Thomas Rutter was a blacksmith, and Jacob Duk, from a family of European armorers, listed his profession as gunsmith. Many of the owners of the new furnaces and forges were merchants, attracted to the investment potential of ironworks.[45] To build and operate their works, they sometimes attempted to induce skilled workmen to emigrate from Europe. More often, they drew upon the expertise already present in the persons of blacksmiths and other metals craftsmen throughout the colonies.

After the close of Hammersmith and before the developments of the early 1700s, the standard method of producing iron was the bloomery forge, the reversion to a practice antedating the development of the blast furnace by many centuries. Any competent blacksmith could

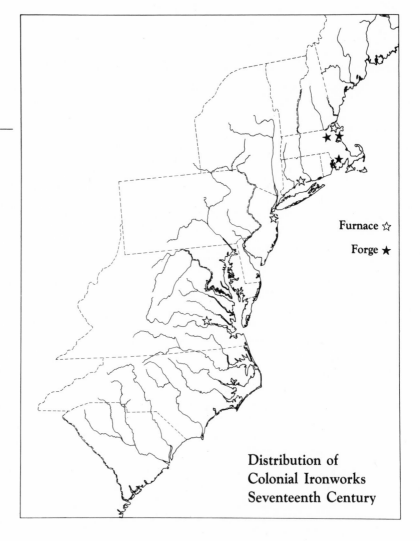

Furnace ☆

Forge ★

Distribution of
Colonial Ironworks
Seventeenth Century

make test of a sample of iron ore and produce small quantities at his forge. A mass of iron ore was heated on a hearth in the reducing atmosphere of a charcoal fire, a process that reduced the metal oxide to iron, leaving a spongy mass of metal and refractory slag called a bloom. Repeated hammering of the red-hot bloom squeezed out the slag and consolidated the iron. Because of the limitations of hearth size and the problem of handling and hammering hot metal, the output of a single bloomery was small. Also, the output nearly always was utilized in the surrounding area so that the expense of overland shipment was avoided. It is probable, however, that many of the more remote towns satisfied much of their requirement for iron from such sources.

Although, as Hartley has shown, the reservoir of skill left from the collapse of the ironworks at Saugus had a prolonged influence on the manufacture of iron in New England, the men who built the first furnaces in the forests of eastern Pennsylvania had no known connection with any previous venture. Therein lay the distinctive feature of the indigenous stage. It was begun and propagated by individuals who had come to America possessing the requisite skills and wished to serve their own interests, rather than those of any government or commercial party. As the iron industry grew in the eighteenth century, it became common to advertise in the newspapers of the major seaports for experienced founders and other ironworkers.[46] It was assumed that the necessary workmen could be obtained directly in America, either men trained abroad who had immigrated or, like the heirs of Hammersmith, men who obtained their technical knowledge and experience by apprenticeship at colonial forges and furnaces.

Although the step from forge to furnace was a large one, as we have seen in the case of Hammersmith and its aftermath, the greatest difference lay not in technology, but in the organizational and business skills required by the complex furnace operation. By the early eighteenth century, those skills also had been developed and had become widespread throughout the colonies. Paul Paskoff has examined the role of merchant involvement in what I have chosen to call the rise of an indigenous American iron industry. Many prominent merchants, such as James Logan of the Durham Iron Company, clearly were attracted to the industry because of the potential for profit. Certainly, well-managed ironworks had long been economically successful in England. There was no reason why careful management could not produce similar results in the colonies. Unlike the seventeenth century, when the primary market was perceived to be in England, and the capital, technology, and managerial skills all had to be imported, there now existed a growing market within the colonies, centered on the port cities of the Northeast, a market that no longer could be satisfied by importing iron from Europe. Sufficient skilled workmen were present to build (if not always to operate successfully) ironworks, and capital generated from the profits of colonial trade and skills honed in the management of that trade were available.[47]

Although its primary purpose was to produce iron for sale in England, the Baltimore Company was built with colonial capital. Within the first six years of operation, the partners invested £14,835 in gold and £2,873 in Maryland currency.[48] In Pennsylvania, the Durham Iron Company listed expenditures of £4,000 in the years 1727 and 1728.[49] Such sums would have been unthinkable for colonial investors in the seventeenth century. Even so, large, well-financed companies were an exception to the apparently more common practice of building upward from forge to furnace by reinvesting profits. Of particular import to the present study, Paskoff also found that, in addition to the quest for profit on the part of the colonial merchant-manufacturers, there was a clearly stated desire to further the public advantage by their endeavors.[50] It would seem to be a long step from aiding the public by alleviating the shortage of a vital material such as iron to winning political independence from England. But, in establishing an iron industry designed to serve the colonial market and thereby lessening the dependence on England for this important commodity, one of the first tentative steps toward achieving total independence was being taken.

In the seventeenth century, the attempt to introduce the full scope of English iron technology proved overly ambitious. The resurgence of the industry after 1718, especially the proliferation of blast furnaces with the characteristic production of cast pig and sow, greatly increased the amount of iron available. Increased production capacity created new problems, including the availability of raw materials, transportation and marketing of output, and control of the furnace technology.

The problem of raw materials was solved with the creation of the iron plantation. For a bloomery operation, local deposits of bog iron, which were plentiful on the eastern seaboard, were sufficient to ensure a steady supply of ore. Bog ore deposits created by the leaching of iron-rich salts from surrounding strata often had the unique advantage of being self-replenishing in approximately a twenty-year cycle, as long as the soil and drainage patterns were not disturbed. The blast furnace, however, capable of continuous operation for months on end, required far greater quantities of iron ore, charcoal, and limestone, the last of which served as a flux to carry away the impurities as slag. The relatively small Braintree furnace erected by John Winthrop had soon exhausted the local bog ore supply and thereafter was used only to refine cast iron from the Hammersmith facility. In Virginia, Colonel Spotswood had to import limestone from England because it was lacking in the vicinity of his furnaces.[51] Among the blast furnaces erected in the colonies during the eighteenth-century boom, some had capacities comparable to the largest charcoal furnaces in England. Warwick Furnace on French Creek in Chester County, Pennsylvania, for example, had an estimated production of 900 tons of cast iron per year[52] and required between 5,000 and 6,000 cords of wood each year while in blast, the product of about 240 acres of woodland.[53] Of the three essential materials, however, the deciding factor in the location of eighteenth-century furnaces was the presence of an adequate supply of ore. To provide the necessary large quantities of ore, charcoal, and flux, it was desirable to situate the furnaces in areas where all these commodities were locally abundant. Extensive surface ore deposits and limestone outcroppings were found in the unbroken forests that began only a few miles inland from the coasts, where running water to power blast and hammer also was available. Furnaces began to locate in the resource-rich hinterlands, giving rise to self-contained communities to which Bining and others have ascribed the term "iron plantation."

A Colonial Blast Furnace
This photograph, made during the reconstruction of the furnace at Hopewell Village National Historic Site, Pennsylvania, shows all but one of the major elements of a typical colonial installation. The massive, square stone furnace contained a hearth arch (foreground), where the molten iron was tapped, and a water-powered bellows unit (right), which provided the blast. Missing here is the walkway to the top of the furnace, through which it was charged. (*Courtesy, United States Department of the Interior, National Park Service*)

The characteristic iron plantation was organized around the demands of an iron furnace and possibly one or more forges. In addition to the ironmaster, who may have occupied a mansion house, a plantation might contain workers' houses, storage houses for charcoal and equipment, an office, a store stocked by the company, where goods were credited to workers often in lieu of wages, a blacksmith shop, barns, and a sawmill. The actual operating staff for a furnace was small—two founders, the principal skilled help who worked in twelve-hour shifts to keep the furnace in continuous operation, moldmakers, "guttermen," fillers, and often a "potter," who was skilled in the manufacture of hollowware. A large part of the plantation population was involved in farming and husbandry to sustain the community. There also were miners for ore and limestone, a large staff of woodcutters, colliers, who converted the wood to charcoal, and teamsters. Land was needed not only for raw materials, especially wood, but for farming and pasture. As a result, the enterprises occupied extensive tracts of land, varying from a few hundred acres for some of the smaller units to such large Pennsylvania plantations as Elizabeth, with 10,124 acres, and Boiling Springs, with 7,000 acres.[54]

Although self-contained, only a few of the very largest units ever achieved self-sufficiency in raw materials and foodstuffs. By far the majority purchased grain and other supplies from local farmers and merchants, employed neighborhood woodsmen, carters, and even blacksmiths for varying periods to supplement their labor force, and traded iron to merchants in the port cities for the dry goods, furniture, clothing, and whiskey they could not produce.[55]

Although the plantation system reached its zenith in Pennsylvania, the pattern was repeated throughout the colonies in the eighteenth century, wherever sufficient quantities of ore were found that might warrant the construction of a furnace. Colonel Spotswood's ironworks complex drew heavily on the wood from a 45,000-acre estate, and Byrd reported that more than one hundred hands, many of them slaves, were needed to support the operation of a single furnace, including women to cook for the men.[56] Later, John Tayloe's Neabsco Furnace in Prince William County, Virginia, occupied a tract of 5,000 acres, and a rival furnace operation attempted by Charles Ewell and his associates owned 1,520 acres.[57] At the start of the Revolution, David Ross noted that the recently built Oxford Furnace holdings contained in excess of 12,000 acres.[58]

The iron plantation flourished in the forested ridges and valleys of eastern Pennsylvania, where swiftly moving streams provided water power and outcroppings of limestone and iron ore were abundant. Nearby Philadelphia was a major market for cast ironware and bar iron. The proximity to market or to convenient means of transportation was an important consideration in the development of eighteenth-century ironworks. Robbins has shown that the extensive tobacco trade with England already existing in the Chesapeake region was a favorable factor in the building of Maryland ironworks. This trade also helped to direct the attention of Maryland ironmasters away from the colonial market toward overseas trade. It is not surprising, then, that during the eighteenth century the major exports of iron from the colonies to England came from Chesapeake ports.[59]

In Pennsylvania and states farther north, ironworks directed their output to the nearest large city for sale or for consignment to more distant markets. Over 50 percent of the ironworks erected in Pennsylvania were within a radius of twenty to forty miles from Philadelphia, connected to the principal market by a road system of highly variable quality. The costs of overland transportation added significantly to the price of iron in the Philadelphia market, averaging about 1.3 shillings per ton mile. To an approximate production cost of £19 per ton for bar iron, transportation charges could add from £3 per ton for nearby furnaces to over £6 per ton for works such as Mary Ann Furnace in remote York County. Thus overland transportation costs confined the early development of the iron industry to a band of less than one hundred miles from the market ports. Beyond that band, the total costs of producing iron and delivering it to market tended, in most instances, to exceed the competitive prices.[60] The concentration of furnaces and forges in eastern Pennsylvania was a consequence, then, both of the availability of raw materials and the economics of overland shipping to the largest colonial market in Philadelphia. When Thomas Rutter and Samuel Nutt built their forges and furnaces to sell to that market, they were beginning the era of the iron plantation in Pennsylvania. They also were filling a need becoming more apparent in the colonies—the need for American sources for iron tools and utensils.

Possessing none of the impressive features of the more famous iron complexes such as Hammersmith, these independent and little-noted furnaces continued to multiply in the third and fourth decades of the century. The Schuylkill River and its tributaries became the locations of the greatest concentration of ironworks in the colonies, with more than fifty furnaces and forges built before the Revolution.[61] Within a few years, from the furnaces and forges of Pennsylvania a steady stream of pots, pans, firebacks, and implements began to move to the colonial market.[62] Bar iron from the plantation furnaces and forges supplied the smiths located in nearly every community. Inspired by the success of these works, the other colonies began actively to promote ironworks within their borders. New furnaces and forges were erected from Massachusetts to Virginia. Subsequently, their combined impact on the colonial market drew the attention of the British government to the economic significance of iron in America. In the building of those furnaces, with their peculiarly American orientation, can also be found the origins of a truly indigenous American iron industry.

5

Metals Manufacture in the Colonial Period

I hope I may be pardoned, if I declare my opinion to be, that all these Colloneys, which are but twigs belonging to the main Tree (England), ought to be kept intirely dependent on and subservient to England, and that can never be if they are suffered to goe on in the notions they have, that as they are Englishmen, soe they may set up the same Manufactures here, as people may doe in England.—Lord Cornbury, 1705[1]

When considering the types of labor to be practiced by his mythical Utopians, Sir Thomas More had concluded that one of the most desirable and essential was the art of metalworking. Metal artifacts were and are an integral part of Western culture. Initially, colonists had no choice but to carry with them to America or to purchase from abroad, primarily from England, the broad range of material goods to which they were accustomed. But the growing population of the colonies came to include native-born Americans and large numbers of people who had paid for their passage to the colonies by indentured servitude. Many of them had little money to expend on imports, but they, too, required a supply of artifacts to participate in the mainstream of colonial living. That any effort to recreate the cultural dimension of European civilization made the production of such artifacts inevitable and necessary was soon as apparent to the American colonists in the seventeenth century as it had been to More one hundred years earlier. Second only to heavy industry as characterized by iron refining was the need for artisans, prominent among them metals craftsmen. Coincident with the rise of an indigenous colonial iron industry in the eighteenth century came the proliferation of metals craftsmen, many of whom directly benefited from the local availability of supplies of iron.

In the eighteenth century, the products of the iron industry were the most conspicuous evidence of the practice of metal technologies in colonial America. Before the output of the furnaces and commercial forges could be turned to everyday use, however, additional processing often was necessary. The conversion of primary metal forms such as castings, rough forgings, bar, and plate into finished products by the secondary operations of hammering, turning, and joining was the work of the metals craftsman.

During the early years of colonization, the pressing demands for food, fuel, and shelter were paramount. The ax and the hoe, the gun and the hammer, products of the blacksmith's art, were the essential tools for the settlers. As Carl Bridenbaugh has noted, it was by necessity the century of the farmer.[2] The influx of colonists did include many men skilled in the production of metal artifacts, particularly in the New England region, but with few exceptions, the chief one blacksmithing, the practice of those skills languished in the seventeenth century. Not until the eighteenth century, bolstered by the achievement of a degree of security and affluence, the demands of a rising population, and the opportunities afforded by the concentrated markets of a few urban centers, were the colonial craftsmen in metals and other trades able to achieve any degree of success or prosperity. Blacksmithing remained the most important and widely practiced form of metalwork, but in addition to "general" blacksmithing a host of specialized ironcrafts such as wheelwrighting and locksmithing began to be practiced. In addition, the manufacture of artifacts from silver, copper, pewter, and tin became established in the colonies.

A comprehensive study of the origins and growth of metalworking trades in North America would carry the present survey far beyond its intended limits. The significance of those trades to the development of metals in colonial America cannot be overstated, however. In particular, the rise of the colonial metalworker had far-reaching consequences for the American colonies: in helping to further economic independence from England, in promoting trade and commerce both within the colonies and abroad, and in creating a reservoir of skills that contributed to the revolutionary cause.

The probated estate records of the colonial courts provide ample evidence of the importance of manufactured metal goods to the early colonists. As far as possible, they attempted to emulate the life style they had left behind in England. To do this, they had provided themselves with the utensils and tools they deemed indispensable. In the seventeenth century, the only source for many of those artifacts was England. It is not surprising therefore that probate records contained detailed listings of clothing, household furnishings, and tools made the more valuable by their scarcity in the absence of colonial manufacturing. Beginning with the studies of Robert Bruce early in the present century, historians have looked to such records for the clues the artifacts provide about social customs, standards, and cultural attainments in the colonial period.

What were the metals the colonists themselves considered important, and in what forms were they used? Bruce found that in Virginia kitchenware and eating utensils constituted perhaps the major category of metal goods in the colonial home. In the kitchen were found iron fireplace racks and andirons, large iron pots, copper or brass boilers, brass, tin, and copper kettles, ladles, and skillets, and steel knives.

Copper and tin were used for sifters, skimmers, saucepans, graters, and funnels, brass and bellmetal bronze for mortars and pestles, scales and weights. Iron and copper and its alloys, because of their great durability, were the metals preferred for items of daily utility. Pewter was the fashionably common metal used for plates, spoons, bowls, saltcellars, cups, tankards, and candlesticks. Spoons often were made of hardened tin or an alloy of brass called "alchemy."[3] More expensive, and hence far less common, was silver. Some silver plate was owned by the more wealthy planters, but in general silver appeared more often in the form of spoons, candlesticks, and decorative tableware. A typical holding of silver among the Virginia gentry is represented by the personalty of William Kendall of Northampton which contained "twenty-seven spoons, four salt-cellars, two sugar-dishes, a porringer, a tankard, two dram cups, two punch and one caudle, and a pair of snuffers."[4] Gold, the metal so eagerly sought by the early colonists, was conspicuously absent from the lists.

The pattern of metal household items found by Bruce for Virginia prevailed throughout the colonies. In his demographic study of family life in the Plymouth Colony, John Demos noted the same general usage of metals by the descendants of the Pilgrims. Kitchenware and eating utensils were of iron, brass, and pewter, although earthenware and wooden plates and bowls were common among the poorest group. Both Bruce and Demos noted the near absence of the table fork in the inventories. Silver was used by only the more wealthy families.[5]

The probate records of Essex County, Massachusetts, just to the north of Plymouth, present a comprehensive picture of the possessions of the Massachusetts Bay colonists. Occasionally they also include some relative values for individual objects. A typical list of metal household goods taken from the estate of John Bartoll of Marblehead could have been found in homes from New Hampshire to the Carolinas.[6]

a great Copper		3 li.	10 s.
one Iron pott			
an Iron ketle			
2 bras skilletts			
one Iron scillott			
and towoe brass scillots	total	3 li.	
5 pewter platters			
and a bason	total	1 li.	
pewter			16 s.
a morter			
and a bras skillet	total		8 s.
2 dripin pans			6 s.
a bandsaw			1 s. 6 d.

Once again, in the Massachusetts records iron appeared most often in the form of fireplace hardware, in pots, frying pans, skillets, and smoothing irons, generally heavy items, many of which could be formed by casting. Copper or brass, more malleable than iron, more often appeared in shapes that required some degree of mechanical working to form, such as candlesticks, kettles, and cups. Pewter, again, was the most common household metal used for plates, spoons, and a wide variety of containers from bowls to chamber pots. Tin vessels were found in only a few inventories but, like pewter, seem to have had the lowest relative value of the metals used.[7]

Characteristic differences in the inventories indicate the occupational needs of the settlers. Thus, Thomas Flint of Salem, who ran a sizable farm, left plow chains, an iron harrow, several plow irons, hoes, axes, wedges, and a variety of hand tools, all of iron or steel. William Odry of Ipswich, a fisherman by occupation, left a collection of lines, iron hooks, lead weights, and sinkers. Iron hammers, saws, augers, chisels, and other carpenter's tools were common items in the lists, as in the estate record of Francis Plummer of Newbury.[8] Where muskets are inventoried in the estates, quantities of lead shot or of bar lead suitable for making shot often are noted, usually the only mention of lead.

One interesting finding was the great value attached to nails. The majority of houses were built without nails because without ironworks in the colonies they could be obtained only from England and at an expense that precluded their use by the majority of settlers. Estate inventories often gave an exact count of the nails in a man's possession. For example, at his death John Carter of Lancaster, Virginia, left the unusually high quantity of "over seven thousand eight-penny, twelve thousand two hundred and thirty-three ten-penny, and nearly five thousand twenty-penny nails."[9] Amounts as small as a few dozen nails were recorded. The sheer volume of metal objects and the great variety attest to the importance of metal to the colonial family.

From the survey above, a clear pattern of the significance of metals to the colonies in America emerges. Iron rated first in importance because of its widespread application in tools of all kinds, both as wrought iron and steel, the latter used for a variety of cutting instruments, and its use in pots and irons and many other important household items. Brass and copper were found almost exclusively in the home in a range of useful artifacts, such as cookware, as was the less expensive pewter. Smaller quantities of tin and silver were used in lamps, candlesticks, and similar domestic objects, the latter mostly in decorative forms. Of course, tin was present as an alloy constituent of both bronze and pewter, with copper in the first instance and with lead in the second. Lead itself was needed for weights in commerce and fishing sinkers but primarily for musket shot. The roster of metals employed in America as shown in the seventeenth-century estate records was essentially the same as that used for centuries in Europe, except that gold was missing. With the exception of iron, none of the metals used in the colonies had been produced there by the start of the last quarter of the seventeenth century. Artisans already were at work, however, especially in the seaport cities, fabricating the metal artifacts needed and desired by the colonists.

Foremost among the colonial metals trades was that of the blacksmith. The importance of iron in all phases of colonial life, from kitchen hardware to rifles and agricultural implements, ensured a steady occupation for the blacksmith from the earliest days of settlement in both North and South and often long before other metals trades, supplying less essential commodities, could achieve a firm footing. Before the end of the eighteenth century, it is relatively certain that any community of modest size had at least one man who practiced the trade. One of the first colonists at Jamestown was James Read, blacksmith, who landed in the summer of 1607 to practice his craft using bar iron shipped from England.[10] In New England, John Robinson at Haverhill (1640) and William Cheseborough at Stonington (1650) began to practice blacksmithing almost at the start of settlement. The establishment of forges suitable for both blacksmithing and the smelting of small quantities of ore to supply raw material often was sufficient to ensure the growth of settlements.[11] Such was the case with the forge built by Joseph Jencks at Pawtucket, Rhode Island, and the several forges of the Leonard family at Taunton, Massachusetts.[12]

In the cities, North and South, were the largest numbers of blacksmiths' shops catering to the general and specialized demands of home and commerce. Although exact figures for the number of blacksmiths in the colonies at any given time are not available, some evidence can be deduced of the number of men who practiced blacksmithing in the urban centers. After a survey of the early documents in the New York City Hall of Public Records, Albert H. Sonn produced the following roster of men who, over a two-decade span, listed blacksmith as their occupation:

> Samuel Phillips, Dec. 5, 1695; Burker Myndearts, Jan. 23, 1697; John Cooley, Aug. 23, 1698; Johannes Van Voorst, Aug. 23, 1698; John Peterson Melott, Sept. 6, 1698; Hendrik Bush, Sept. 6, 1698; Abram Van Aarnan, Sept. 6, 1698; Martin Beeckman, Sept. 6, 1698; John Breesteade, Feb. 3, 1699; John Bachan, Mar. 27, 1700; Joseph Hart, May 27, 1702; Richard Steward, May 27, 1702; Thomas Hews, May 27, 1702; William Hall, May 30, 1702; Andrew Hannis, Aug. 5, 1707; William Brown, Oct. 11, 1710; William Bouquet, Apr. 14, 1713; Thomas Elder, Apr. 24, 1716; Ebenezer Mors, Apr. 24, 1716; Abraham Price, June 4, 1717.[13]

This was during a period when the population of New York had not yet reached five thousand. A 1774 tax list for Philadelphia recorded fifty-two master smiths for that city alone, all but a few of whom were workers in iron.[14] Yet the colonial blacksmith, the most common practitioner of the metal trades during the period, remains the most anonymous of the many workers in metals. Colonial newspapers carried numerous advertisements for silversmiths, pewterers, bell founders, and other tradesmen, but the number of advertisements relating to black-smithing were, by comparison, few.[15]

The relative anonymity of the colonial blacksmith stems in part from the very number of men who practiced the trade, together with the extensive volume and nature of their products—practical, inexpensive, and aesthetically plain. In the wide-ranging rural areas away from towns and along the frontier, such simpler skills of the smith as the making of nails from slit bar stock were practiced by the individual farmer or settler. Much smith work, such as reedging tools, was in the line of maintenance rather than manufacture. The occasional special-ists in decorative ironwork often obtained a widespread reputation, men such as Tunis Tebout and William Johnson of Charleston, South Carolina, who were engaged in the early production of the ornamental grills, fences, and balustrades for which the city was noted even in colonial times. Such men were the exception, however. The majority of blacksmiths and ironworkers concentrated on objects of high utility rather than expensive decorative pieces.[16]

The basic tools of the blacksmith had not changed over time, and, essentially, they remain the same today. At the heart of the operation was the forge, a hearth on which a characoal fire burned, fanned to a high heat by forced draft, using a hand-operated bellows. Into the bed of the coals the blacksmith pushed the iron to be formed, and he worked the bellows until the iron began to glow. Once it reached the proper temperature, determined by the color of the metal and long experience, the iron was seized with tongs and taken to the anvil where it was beaten using hammers of various weights and shapes. Different portions of the anvil were used to create a wide variety of forms. The heating and hammering would be repeated many times until the desired configuration was achieved. A final step might be to "draw" the iron or temper it to give it toughness by plunging it into a bucket of water or brine.[17]

Nearly all of the blacksmith's tools could be made by the smith himself from bar iron stock. Throughout most of the seventeenth century, the blacksmith had to depend on imported iron for his raw material. In New England, first Hammersmith and then the local forges that grew up in its wake provided a source of material for local smiths. After the early eighteenth century, except in the southernmost colo-nies, iron from domestic furnaces or commercial forges was widely available. In some shops, the forges were adequate to convert small quantities of local ore to iron, and it was common for the blacksmith to maintain a large junk pile of old iron that supplied the starting material for much of his work. The versatility exhibited in reusing materials was a characteristic the blacksmith shared with all other colonial metal-workers.[18] Only in one area did the smith remain dependent on imports. His cast anvils, weighing from 150 to about 300 pounds, were made in England until well into the eighteenth century. Eventually those, too, could be obtained in America. During the Revolution, blacksmiths' anvils were a common product of the air furnaces erected in many cities to supply local needs.[19]

The blacksmith was able to produce materials in an endlessly varied assortment of sizes, finishes, and properties using dexterity, finesse, and a true knowledge of the nature of his materials acquired by long practice and experience, starting with an extended apprenticeship to a master craftsman. Indenture records for Philadelphia in the mid-eighteenth century indicate that a minimum apprenticeship for grown men would be about five years, while for children indentures as long as nineteen years are listed.[20] Apprenticeships were far from easy work, and colonial papers frequently contained notices offering rewards for the return of boys who had found the discipline and long period too confining, or of older men, qualified smiths, serving out indentures to pay for their passage to America.[21]

In both town and country, a staple of the blacksmith's trade was domestic hardware. Door hinges, latches, knockers, and simple locks were common items. Although the emphasis was on function and low cost, the design and skill of execution evidenced the pride of workmanship that characterized all of the colonial metal crafts, so that many colonial patterns still are copied today, produced by modern machinery. Within the house, the kitchen often exhibited a wide range of the blacksmith's art. Andirons, pothooks, cranes, shovels, fire tongs, trivets, and skewers of iron were present in most homes. Wrought iron candlesticks and betty lamps also were popular. Because such objects received hard use, repair and replacement were common, representing an important part of the blacksmith's job of supplying "essential" items to the households of a growing population.[22]

The agriculturally oriented Tidewater colonies, from Maryland southward, possessed few urban centers and, in consequence, developed a different pattern in the practice of blacksmithing. Many of the needs of plantations accessible via the broad network of bays and rivers could and indeed often had to be met by the import of manufactured goods from England or the northern colonies. Shipping records for Port Roanoke, North Carolina, a colony blessed with few natural resources in her more populous though thinly settled eastern counties, showed substantial imports of bar iron and ironware as late as 1775. From Bristol in England came a steady shipment of nails and castings, from Glasgow, thousands of pounds of castings and bar iron. New England ports and New York were the principal sources for axes and cast hollowware.[23] The lack of urban centers to provide a ready market and the problem of raw material supply tended to limit the practice of independent blacksmithing in the South. In the eighteenth century, it became common to train indentured servants and, later, slaves to do general blacksmithing for the plantations and their smaller neighbors using imported bar and plate stock.[24]

In examining the distribution of colonial craftsmen, Carl Bridenbaugh has suggested that the metal artisan was absent through-

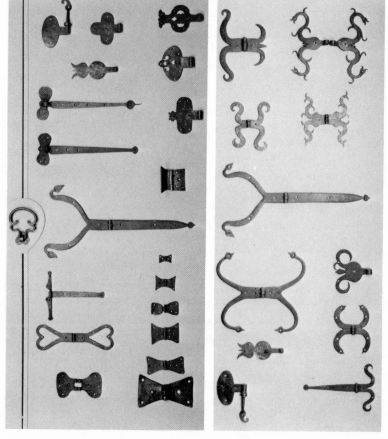

Colonial Iron Hardware
The products of the colonial blacksmith were essential to every phase of colonial life. As this display of iron hardware for doors shows, those products were functional, durable, and aesthetically pleasing. (*Courtesy, The Henry Francis du Pont Winterthur Museum*)

out much of the South because the planters preferred imported goods that "were both better and cheaper" than any that could be produced locally and because most crafts were carried on by slaves against whom the independent craftsman could not compete. Both arguments, however, appear open to question. An examination of surviving articles of colonial manufacture immediately dispels any claim to the superiority of imported goods. Colonial smiths possessed skills equal to their British counterparts. European styles were meticulously copied where tastes so demanded. If price alone was the determining factor, it should have operated more extensively in the port cities of the North, where frequent trade with Britain was conducted. Since smiths prospered in those cities, it appears that they were able to compete successfully in terms of both quality and price. Yet against such evidence one cannot discount the established habits and bias of the Tidewater societies, which were tied closely to Great Britain by trade patterns, as contributing to a scarcity of metals craftsmen.

The second argument, however, is commonly offered to account for the sectional differences in manufacturing that developed in the nineteenth century. The present study suggests that the use of slaves to perform the work of skilled craftsmen may have originated because the rural dispersion of population, creating problems of supply and market, did not attract the independent craftsman. Where a ready market existed, for ornamental grillwork in Charleston or for silversmithing throughout the South, the independent metalworker was found. Among the metals crafts, it is interesting to note that the blacksmith suffered most from slave competition, the one craft most indispensable yet most vulnerable because of transport problems.[25] More study will be needed to resolve this issue.

Although most urban and rural blacksmiths produced a wide range of articles on individual order a number of men engaged in specialized lines of work. Among the latter were the makers of decorative grates and grills in the larger cities like Philadelphia and Charleston. (The more well-known wrought iron balustrades of New Orleans date from the early nineteenth century.) Unlike the general blacksmith, who had learned the rudiments of his trade during a long apprenticeship and developed new patterns and shapes by imagination or improvisational necessity, the makers of ornamental ironwork often had recourse to "pattern" books or samples of the latest fashion in wrought iron forms developed in Europe. One such pattern book, popular in Charleston, South Carolina, just before the Revolution, had been printed in London in 1765 under the title, The Smith's Right Hand or a Complete Guide to the various Branches of all sorts of Iron Work Divided into Three Parts.[26] Study of the transfer of metal technologies from Europe to America and their subsequent growth in the colonies yields very little evidence for the influence of a technical literature in the process. The use of pattern books for ornamental grillwork and the emulation of European styles in silverware often copied from illustrated books constitute the main literary influence on colonial metalworking or manufacture.

Locksmithing, the forerunner of a nineteenth-century industry, was primarily an urban specialty, although usually combined with ironwork in other areas and even with other metals. The locksmith and another urban craftsman, the cutler, both suffered more than other ironsmiths from competition with imported English goods. This was especially true for cutlers because the steel necessary to produce their stock-in-trade had to be imported throughout nearly all of the colonial period. Hence most cutlers carried a full line of imported edgeware which they ground and serviced as part of their trade. The cutler's stock included edge tools, such as adzes, axes and chisels, and knives, razors, scissors, and surgeons' instruments. Edge tools, like axes, had a wedge of steel welded onto the iron head. After repeated sharpenings, the cutler could reedge the tool by replacing the worn-out blade with a new "steel." Only in the decade before the Revolution was steel beginning to be produced in quantity near the urban centers of Philadelphia and New York. The production was devoted primarily to edge tool manufacture. Small items, like razors and surgical implements, were of single pieces of steel. These continued to be imported and advertised as such right up to the Revolution.[27]

The general blacksmith, with his forge and anvil, could be found in every small town throughout the colonies, but one highly specialized branch of the art—whitesmithing, which involved filing, polishing, and assembling of iron parts as its principal operations—was largely confined to the cities, where finely finished articles found a ready market. Robert Anderson of New York was typical of the colonial whitesmith, advertising ash pails, smoke jacks, and scale beams as part of his stock.[28]

Finally, there remains one class of specialized blacksmith in the coastal towns and cities about which all too little is known, the shipsmith. Early in the seventeenth century, the building of ships had commenced in the northern colonies, and by 1724 ships of up to one hundred tons displacement were being turned out of yards from Maine to the Chesapeake, with numerous smaller craft, suitable for navigating the shallow coastal waters, being made farther south.[29] Throughout much of the early period, any iron used in ship construction was imported because none was available from the colonies. Wood was substituted for iron wherever possible, particularly in joining heavy members where shaped oak pins, or treenails, were used. Even so, the smith performed a double function in shipbuilding.

The first function was in the supply and maintenance of tools. Among the more than forty types of tools employed by shipwrights were up to ten sizes of augers, frame, cross-cut, and whip saws, hatchets, hammers, cold chisels, scrapers, planes, and caulking irons.[30] Initially, most such tools were imported, but by the beginning of the eighteenth century many could be supplied and all repaired by local blacksmiths.

The second function pertained to the construction itself, where bolts for scarfing keels and beams and nails to fasten planking to them were cut and shaped on the spot. It is estimated that a ship of one hundred tons displacement would need one ton of ironwork. The cost of iron was roughly 16 percent of the total for a colonial vessel, making it one of the major expenditures.[31] Until well into the eighteenth century, few forges for making anchors existed in the colonies.[32] For smaller vessels, the smith could shape an anchor from bar stock. Large anchors had to be imported.

There is at least one record of blacksmiths undertaking the complete task of building ships. John Barker and his three sons moved to Maine from Hanover, Massachusetts, where they had listed their trade as blacksmith. In 1728 they launched their first ship near the village of Dresden Mills.[33] Despite the important role played by smiths in the colonial shipbuilding industry, few shipsmiths are known to us by name. One exception is Jacob Whitman of Providence, who specialized in the manufacture of ship hardware. Whitman's well-frequented shop became known as a city landmark in the 1750s.[34]

Specialized blacksmithing was far less common in rural areas. A survey of the craftsmen in Bucks County, Pennsylvania, in the second half of the eighteenth century lists eighty-nine men classed as "blacksmiths" and thirty-six as wheelwrights. For specialized work, only eight men were noted in a fifty-year span—four nailers, one cutler, a shipwright, and two farm tool makers. Surprisingly, the trade of farrier, the main activity of which was the shoeing of horses, was not practiced as a separate occupation in largely rural Bucks County. Because of the lack of hard-surfaced roads in the colonies, it was not so necessary to shoe horses to protect their hooves, so that farriery was not common before the nineteenth century.[35]

Compared to other skilled trades, the blacksmith enjoyed a slightly higher degree of prosperity. In 1714, Morgan Evan of Pennsylvania wrote to a friend in England that, whereas a good carpenter was paid 2s. 6d. per day and a tailor 1s. 6d., a blacksmith could receive 3s. for shoeing a horse or installing plow irons or 12s. for the irons and 6s. for a pickax if he made them from his own stock.[36] Evan was writing at a time when colonial iron still was scarce. The availability of iron from local sources offset the higher costs that would have resulted from increasing tariffs imposed by England on iron exported to America. The daybook of a blacksmith near Colebrookdale, Pennsylvania, showed that as late as 1760 the cost of shoeing a horse remained at 3s., while the charge for a shovel was 4s. 6d. and for a grubbing hoe 7s. 8d.[37]

One iron craftsman who flourished in the rural areas was the wheelwright. The "ironing" of wagons included making such parts as chains, brake levers, bands for hubs, iron "tires" for wheels, and a variety of fasteners and fittings. The Lancaster-Carlisle area in Pennsylvania had a number of such craftsmen. It was in Lancaster that the Conestoga wagon was invented near the banks of a creek with the same name.[38] There also was the northern terminus of the Wagon Road which, in the eighteenth century, carried an increasing stream of settlers southwestward along the flanks of the Appalachians into the Piedmont of the Carolinas and Georgia. A survey of wheelwrights in North Carolina indicates that up to 1770 the numbers of such craftsmen were small and confined to the eastern third of the state. Between 1770 and 1780, however, at least ten wheelwrights are listed for Mecklenburg and Rowan counties in the west alone, these counties lying on the route of the Wagon Road extending southward from Philadelphia and Lancaster. An entire region was settled and supplied by wagon, and the wheelwright was nearly as important to that vast area as the general blacksmith.[39]

Lancaster was also noted for the number of men who practiced the trade of gunsmith there. In the seventeenth century, the wilderness was conquered with the gun and the ax. The need for firearms became more urgent in the eighteenth century, as friction with the Indians increased along the frontiers to the north and west and sporadic wars with the French threatened old and new settlements. Throughout much of the colonial period, the standard firearm for soldiers and many citizens was the smoothbore musket. By 1700, the flintlock or fusil largely had superseded the matchlock, but the original name was retained to signify an unrifled weapon.

In eighteenth-century England, the Birmingham area became the center for the manufacture of barrels, locks, and other iron and brass hardware. Many of the supplemental parts were cast in sand molds. Both fully assembled weapons and parts were imported to America, the gunsmith carefully filing and fitting the barrel, lock, and plates to wooden stocks. The musket was relatively easy to assemble and maintain so that by midcentury every city of any size in the North and many in the South had one or more gunsmiths to sell and repair weapons. Typical was Maryland, where a survey of gunsmiths at the beginning of the Revolution listed three at Baltimore, four at Fredericktown, one near Frederick, two at Hagerstown, one at Jerusalem, and "several" others on the eastern shore.[40]

In the Lancaster area evolved the long rifle, also called the Pennsylvania rifle, one of the unique developments of the colonial craftsman. It was the response to a need for a weapon with longer range and greater accuracy than the standard musket for use in the heavy forests of the frontier. Evolved from traditional forms of European firearms, it was distinctive because of the long barrel (up to five feet, compared to four for a musket). The availability of ample bar iron from Pennsylvania furnaces and forges made possible domestic manufacture of nearly all components of the long rifle. One exception was the lock mechanisms, usually imported before the Revolution, although these, too, were being cast locally in some quantity by 1775.

In the forging and boring of the barrel the skill of the smith was most vital. The barrel was made by forging and welding a long bar, or skelp, into a narrow cylinder. A portion of the skelp was heated to a white heat at the forge and then hammered around a mandrel, or bick iron, using a grooved anvil. Repeated heating and hammering was necessary to shape the cylinder, which then was bored to a uniform diameter by a hand-operated boring engine. A similarly constructed rifling engine using an indexing device to advance the cutters cut the grooves that gave the rifle its characteristic accuracy. The barrel was finished by grinding to a round or octagonal shape, before being tapped to insert the breech. The main tasks of the gunsmith were to make the barrel and to assemble the parts. As surviving samples of the product demonstrate, his skill was on a par with the finest efforts of his European counterparts.[41]

By the mid-eighteenth century, the skills of the smith were scarcely less essential than those of the farmer and ironmaster to the mainstream of a growing and prospering society in the colonies. In April of 1775, those skills acquired a new and unforeseen importance that will be explored in more detail in a later chapter.

When one considers the repeated claims of historians that throughout the colonial period the colonies were chronically afflicted with a shortage of specie and with a constant and sizable debt to British merchants, it is surprising to discover that, second to blacksmithing, the metal craft most widely practiced was silversmithing. The evidence for these assertions is overwhelming. How, then, can one account for the popularity of silversmithing in the colonies? The most plausible answer is found in the unusual roles of silver plate and the silversmith in the colonial period.

For the colonist, the possession of silver plate had a double significance. On the one hand, it was a reflection of affluence and gracious living, enhancing the social prestige of the owner. Therefore it was especially popular in the northern urban centers among the families of merchants and the professional classes. Silver plate also found a ready market in the more rural South in the homes of the large planters of the Tidewater region.[42] Then as now, however, the distribution of wealth in society was such that the demand for silver coming from the more affluent families would not, in itself, have supported an extensive silversmithing trade. Families of middle or lower incomes also purchased silver plate for a second and uniquely colonial reason.

The second aspect of the ownership of silver was that it provided a convenient means of keeping specie in the days before banks were available. Many modest households held a few silver spoons or small pieces of plate stamped with the maker's mark and personalized with the owner's initials. Within the colonial economy, plate also served as money in the amount equivalent to the silver it contained. By converting coin to more easily identifiable plate, the chance of loss through theft, which was common enough, was diminished and the chance of recovery brighter. Consequently, the newspapers of the colonial period often contained advertisements relative to the theft of "worked" silver items. The following notice in the *Boston Evening Post* for November 27, 1763, is a typical example: "Two dollars Reward, Lost or Stollen, two oldfashioned Silver Spoons, Marked NG–EG. If returned there will be no Questions asked. N. B. The Handles are likewise mark'd downwards IS. IS. N&c. Notice should be given to the Printers."[43] In that sense, the smith functioned as a banker as well as a craftsman. Both factors, but particularly the latter, led to the widespread practice of silversmithing in the colonies in the eighteenth century.

The production of silver artifacts in the colonial period has drawn more attention than any other phase of early metal manufacture. Part of the reason is that a higher proportion of silver items have survived because higher cost and decorative functions made them less susceptible to destruction through constant usage. Many books have been published describing the collections of early American silver in public and private museums, and monographs have been written describing the craft in nearly every state where the manufacture of silver was common in the colonial period. As a consequence, the colonial silversmith is the best known of all of the colonial metals craftsmen.[44]

In the seventeenth century, Boston and New York were the chief centers of silversmithing in the colonies. As the principal port sites, they had greatest access to a supply of metal from abroad, chiefly English coins and Spanish dollars coming from trade with the mother country and the south of Europe. The development of silversmithing in Massachusetts is especially well documented. Included in the great migration of English Puritans who came to Massachusetts before 1640 to establish permanent homes were many families of above average means who brought with them substantial amounts of silver plate. In the next decade, much silver was sent as gifts from England or was purchased there by the colonists for personal use or to adorn the growing number of churches in the Bay Colony. As a consequence of the wide use and need to repair silver, the art of silversmithing early began to be practiced in the colony by European-trained craftsmen.[45]

Enough silver was present in Massachusetts so that by 1652 the General Court voted to establish a mint in Boston to produce coins both for the colony's use and to discourage the influx of counterfeit and depreciated currency that was plaguing the business community. The court selected as mintmaster John Hull, a twenty-eight-year-old native, who had learned the trade of silversmith from his London-trained half-brother, Robert Storer.[46] Hull chose as his partner Robert Sanderson, who had served a nine-year apprenticeship in London, and began the production of a series of coins made with dies cut by Joseph Jencks, who already has been noted for his connection with Hammersmith.[47] On the instructions of the court, Hull and Sanderson maintained the sterling standard for their coins (minimum 92.5 percent silver content). Many of their coins used the motif of a towering white pine, a product of the New England forest prized for use as ships' masts, hence their designation by modern collectors as pine tree shillings.

Metals Manufacture

Hull and Sanderson also maintained a partnership to produce silverware, and many of the coins they made at the mint were returned to their shop for melting into plate.[48] The mint operated until 1688, when it was closed at the demand of the English government. Since minting was the exclusive prerogative of the crown, Massachusetts had acted in open violation of English law.[49]

Silversmithing in New York began under the Dutch prior to 1650, the principal purchasers of plate and hollowware being the wealthy patroon families whose estates stretched northward along the Hudson. In the latter half of the century, the still predominantly Dutch community of silversmiths was reinforced by an influx of French Protestants fleeing persecution after the revocation of the Edict of Nantes in 1685. They adopted Dutch styles in silver, which prevailed in New York until well into the eighteenth century. The Dutch-Huguenot influence on New York silversmithing is illustrated by the career of Bartholomew Le Roux. Arriving in New York via London shortly before 1688, he was to become the best known of the Huguenot silversmiths. His son, Charles, whom he trained, obtained such proficiency that he was made official silversmith to the city, while an apprentice of Dutch ancestry, Peter Van Dyck, Bartholomew's son-in-law, became recognized as the finest smith in the colony.[50]

Silversmithing spread rapidly throughout the colonies in the eighteenth century. Maryland and Virginia had craftsmen practicing the trade as early as 1668[51] and 1694,[52] respectively, and both saw rapid increases in the number of silversmiths in the eighteenth century. As in the northern colonies, they concentrated in the seaports of the Tidewater region. In Virginia, for example, the centers of the silversmithing trade were the towns of Norfolk, Williamsburg, Alexandria, and Fredericksburg.[53] North Carolina records indicate that only thirty-four silversmiths practiced in the colony during the entire eighteenth century, but nine of them were located in Edenton, a port city on Albemarle Sound.[54] Charleston, South Carolina, had at least one silversmith by 1696 to cater to its merchant and planter classes. The number in the colony grew to eight in 1725 and at least twenty-three by 1750.[55] The practice of silversmithing in Pennsylvania began shortly before 1700, but, coincident with the growth of Philadelphia into the largest city in the colony before the Revolution, the colony also became the center of the trade in the eighteenth century.

The practice of silversmithing in the colonial period was oriented to the peculiar needs of the society. The colonial customer would save silver coins until a sufficient quantity had accumulated and then take them to the smith. Once the coins had been weighed and an agreement on the number and kind of pieces to be made had been reached, the smith would melt the coins in a small furnace. The first step was to refine the silver. Additional copper then was added to the liquid silver, forming an alloy with greater strength, before casting into ingot molds. Allowance was made to the customer only for the amount of silver actually used. The ingot was beaten and forged on a heavy anvil to the desired thickness and a pattern for the final product engraved and cut out. To finish a piece, the metal was hammered cold, but, since silver work-hardens, frequent annealing to soften the metal was necessary. A charcoal fire was used, fanned by the blast from a hand bellows. In the finishing operations, if the silversmith used a lathe, it was foot-operated. The smith turned the metal to shape and true cylindrical forms, not to spin form the metal, shaping it by pressing it over a rotating block, which was a later development. Engraving was a hand operation, but the ornamentation common on spoons and flatware was impressed by dies, using either a simple drop press or screw pressure. Intricate items such as teapots were made in several pieces, shaped and/or cast, and soldered together before a final polishing with burnishing stones or pumice powder.[56] To practice the many operations of the trade, the smith maintained a large inventory of hand tools. The estate of Richard Cowpers of Boston, for example, included eleven anvils, sixty-three hammers, and numerous scales, weights, bellows, tongs, shears, punches, and irons valued at nearly £100.[57]

Silversmith Shop, Tools, and Products
To create articles of high quality and timeless beauty, smiths working with silver, copper, and tin required the command of skills that included alloying, casting, forming, joining, and finishing. These skills were acquired through a long apprenticeship and "hands on" experience. (*Colonial Williamsburg Photograph*)

Silversmithing, more than any other metal craft, was responsive to changing style. American tastes tended to reflect those of England, which, in turn, responded to changing fashions on the Continent.

Although designs occasionally originated in printed sources—designs drawn from Dutch books in New York or patterned after the *Analysis of Beauty* by the English painter and silversmith William Hogarth—the colonial styles largely were copied from imported English silver.[58]

Although the first generation of American silversmiths received their apprenticeships in England, France, or Holland, they soon began to teach the skills to a native-born generation. Hull and Sanderson trained two of the best-known colonial silversmiths, Jeremiah Dummer and his brother-in-law, John Coney. Among the second generation were such accomplished craftsmen as Philip Syng, Jr., and Joseph Richardson of Philadelphia.[59] On the basis of their skills, both families rose to prominence in the colony, while continuing the practice of the trade through three generations during the colonial period.

The major stock-in-trade of the colonial silversmith was custom work, but there was a steady trade in buckles, brooches, pins, combs, and other small items of their own manufacture, particularly in the urban areas. One unusual line of goods was the manufacture of jewelry for specific use in the Indian trade initiated in Philadelphia by the elder Philip Syng and Joseph Richardson about 1760 and later carried on by their sons.[60] Silversmiths in the major cities also carried a stock of items imported from England, and they were not above taking advantage of circumstances to promote business. During the crisis over the Townshend Acts in 1768, when the colonies elected to show their displeasure by establishing a boycott against the importation of British goods, the flow of silver also was affected. Simon Coley, a New York silversmith, to the dismay of his competition, chose to ignore the boycott, but an enraged populace forced him to leave the city. On the eve of the Revolution, Charles Bruff, also of New York, advertised a selection of silver and gilt sword hilts available for the men then arming "in Defence of their Liberties."[61]

The price of silver rose throughout the eighteenth century from 7s. per ounce in 1700 to 14s. in 1722, to 34s. in 1744, and to over 50s. per ounce after 1750.[62] The rising price of silver and varying economic conditions in the colonies meant that most silversmiths had to contend with an intermittent market, so they perforce engaged in a variety of related activities including the making of pewterware and the sale of dry goods. Many commonly advertised their talents as engravers or jewelers to make ends meet. Others, such as John Potwine of Hartford, engaged in general merchandising, selling a broad range of goods, including broadcloths, lawns, nutmeg, mace, and imported pewter.[63] Joseph Hopkins of Waterbury practiced law on the side, subsequently ending his career as a judge of probate court.[64] And a few, including perhaps the most famous colonial silversmith, Paul Revere, utilized their skill with metals to practice dentistry.[65]

To the revolutionary struggle and economic unrest of the early national period the silversmiths contributed little of direct consequence. Their presence in colonies, however, was yet another reflection of the growing spirit of economic and cultural independence and the desire on the part of the colonists to procure the artifacts they wanted without the expense or inconvenience of sending to England for them. In that sense, the practice of silversmithing was an integral part of the evolution of metal technologies in the American colonies.

Similar in skills to the silversmith but filling a more utilitarian role in colonial society was the coppersmith. The term coppersmith for a craftsman skilled in the making and repair of copper and copper alloy products had broad connotations in the colonial period, encompassing such specialties as brassworking or founding, for which the specific terms brazier or founder also were used. Like his counterpart in silver, the coppersmith also often engaged in the production of pewter, tin, and leadware. Coppersmithing, however, was slower to develop in America than either black- or silversmithing. The reasons for this may best be expressed in terms of the related problems of supply of raw materials and market for products.

In the seventeenth century, workable copper deposits had not yet been discovered in America. The sole source of supply throughout much of the period was England. Importing sheet or its principal alloys, brass and bronze,[66] was expensive. Although copper had superior forming and heat transmission qualities to iron, for decades many of the items later produced in quantity by colonial coppersmiths were more commonly and cheaply fabricated from readily available iron sheet. The economic considerations acted as an early barrier to the development of the craft. So limited was the practice of coppersmithing in the colonies that, when copper ores were discovered in Connecticut and New Jersey early in the eighteenth century, the ore was shipped to England rather than smelted for local use. Even the building of a copper smelter at Simsbury, described earlier, was for the purpose of conserving fuel in England, not to produce copper for colonial use.

With the rise in population in the early eighteenth century and increasing prosperity in the colonies, a more favorable climate for the skills of the coppersmith evolved and coppersmithing began to develop in the major metropolitan centers. Partly this was necessary because of a dependence on imports for raw materials, but the port cities of Boston, Providence, New York, and Philadelphia also afforded ready markets for more specialized aspects of the craft. Tied closely to the supply of raw materials and large markets, coppersmithing remained concentrated in the northern and urban areas throughout the remainder of the colonial period.[67]

The advertisement of Thomas Russell in the Boston News-Letter, October 30/November 6, 1740, presented a typical picture of the versatility of colonial coppersmithing as practiced in the urban centers: "Thomas Russell, Brazier, near the Draw-Bridge in Boston, Makes, Mends, and New-Tins, all sorts of Braziery ware, VIZ. Kettles, Skillets, Frying-Pans, Kettle-Pots, Sauce Pans, Tea Kettles, Warming Pans, Wash Basins, Skimmers, Ladles, Copper Pots, Copper Funnels, Brass

Scales, Gun Ladles, &c. makes all sorts of Lead Work for Ships, Tobacco Cannisters, Ink Stands, &c. and buys old Brass, Copper, pewter, Lead and Iron."[68]

Russell's stock-in-trade identifies a main contribution of the copper-smith to colonial society—the supply and maintenance of numerous domestic articles, including two of the most essential possessions of a colonial household, the tea kettle and, in cold New England, the warming pan. Although often exhibiting the same standards of work-manship and aesthetic quality that have made colonial silver so valuable, most copper artifacts suffered hard use. As a result, far fewer of them have survived to the present, a fact that contributes to the relative obscurity of the makers.[69]

Russell's list of goods also indicated another element that contrib-uted to the rise of coppersmithing in the eighteenth century. Brewing and distilling were popular and lucrative trades for the colonists almost from the beginning of settlement. Nearly every inn made its own beer, and by the mid-eighteenth century, New England rum had become a staple of the colonial export trade. Because of its resistance to corrosion and high heat-transmitting ability, copper came into demand for the manufacture of large kettles and stills, skimmers, funnels, and ladles.[70] Other emerging trades also used large copper kettles for heating liquids, among them dyeing, hat making, and such diverse occupations as the production of soap and apple butter.[71] Once again, the importance of having skilled metalworkers at hand is apparent. The failure of a still or the lack of kettles in which to brew beer or felt beaverskins for hats would have been more than an annoying inconvenience while replace-ment parts were ordered from abroad. The livelihood of numbers of people and, particularly in the case of rum exports, an important element of colonial trade depended on the supply and repair of copperware. For the colonists, dependence on England for such goods and services was no longer adequate. Economic necessity dictated that, as with a growing supply of iron from American furnaces, those goods and services be present in the colonies.

The basic stock for the coppersmith was copper sheet. The sheet itself was formed from a cast ingot by "battering," using water-powered drop-hammers or, in the eighteenth century, by passing the ingot through water-driven rolls.[72] Although battering may have been used to a limited extent in America, it is certain that the first rolled copper sheets in the colonies were not produced until the eve of the Revolu-tion, in New Jersey, after special rolls had been imported for that purpose.[73] An important use of copper sheet that developed in England during the eighteenth century, the sheathing of ship bottoms to protect against destruction from marine growths, was not adopted in America before the Revolution.[74] The more highly developed technology of casting and rolling copper sheet in England, when coupled with the relatively low demands of the colonial market, best explains why copper continued to be imported even when the ore was being mined in the colonies.

Sheet usually was imported, but a substantial quantity undoubtedly was obtained from the salvage of old copper vessels. In 1771, for example, the northern colonies, the principal market for British copper and brass, imported only twelve tons of copper, mostly in the form of sheet, and twenty-nine tons of brass.[75] After the sheet was cleaned and flattened, it was cut to a pattern and shaped by tools, using techniques similar to those of the silversmith. The requisite skills, especially those of joining the edges of formed sheet to assure a watertight seam or the technique of casting brass to control the loss of volatile zinc and to ensure soundness in the cast pieces, once again were acquired by serving a careful apprenticeship to a master craftsman.

Sheet metalworking, however, was just one aspect of the practice of the colonial coppersmith. Another skill important in the colonies was that of copper foundry work, which also encompassed the casting of brass and bronze artifacts. The practice of casting copper presented different problems from working with the metal in sheet form. Copper, though easily melted, takes on oxygen, in the process becoming viscous, often preventing the filling of the mold. To increase the fluidity of the copper so that it would fill the mold completely, a common practice was to add lead to the melt. Bell and gun metals were made by alloying the copper with 8–12 percent tin, a composition that had been recognized for more than three thousand years as giving the metal its maximum strength.

Brass founding was a specialized art. Up to the eighteenth century, brass was made by adding a mixture of calamine or zinc carbonate and powdered charcoal to a copper bath. Not until 1781 did James Emerson, an Englishman, patent a method for making brass by directly alloying copper and zinc metal.[76] Throughout the eighteenth century, brass was made in the colonies by the old method, occasionally using imported calamine but primarily by remelting scrap. Zinc ores were not mined in America during the colonial period. The direct casting of brass by copper and zinc (also imported) was pioneered in America by the Porter brothers in Waterbury, Connecticut, after 1806 for the making of brass buttons.[77] As with other metal crafts, the basic technology of brass founding was brought to America by men who had learned their skills in England. They in turn passed it on to apprentices so that the number of founders expanded by growth from within as well as by immigration throughout the century. The following articles of indenture are typical of the practice: "Daniel Jones, son of Mary Jones, by consent of his mother, indents himself apprentice to Richard Allen of Phila., brass founder, for six years from Nov. 20th 1745, to be taught the trade of a brass founder and at the expiration of his apprenticeship in case his mother should die before that time, to have a new suit of clothes."[78]

Foundry work produced a variety of useful and decorative items such as andirons, weathervanes, bells of all shapes and sizes, gun, ship, and door hardware, buckles and buttons, and the intricate parts of scientific instruments. One of the best-known products of the colonial founders' art was the casting of the great bell for the State House in Philadelphia, later called the Liberty Bell, by the firm of Pass and Stow in 1753. The history of the bell is illustrative of the difficulties that faced the founder. The bell originally was cast by an English firm but cracked soon after being installed. Such failures were common. It was difficult to control shrinkage and stresses in large castings or to prevent the formation of voids or the inclusion of large impurities in the metal. It took Pass and Stow two attempts to remelt the bell and to obtain a sound casting. Eventually this bell also cracked (in 1836) after years of use.[79]

Not all producers of copper artifacts were skilled craftsmen. One of the first men to use native American copper was Dr. Samuel Higley of Simsbury, Connecticut, who owned and operated a small mine not far from the better-known Newgate lode. From 1737 to 1739, Dr. Higley and his brother, John, attempted to alleviate a shortage of hard money in the colonies by minting a series of coins bearing such colorful mottoes as "I.Am.Good.Copper" and "Value.Me.As.You.Please."[80] It is interesting to note that, though the first copper coins made in America used domestic copper, for decades after independence the copper blanks for coinage continued to be imported from England.[81]

In the eighteenth century, the growth of the colonial shipbuilding industry and a significant increase in maritime trade conducted by the colonies encouraged the spread of yet another specialized form of coppersmithing. This was a demand for the creation of scientific instruments. At sea, the need for telescopes, compasses, dividers, sextants, and other navigation instruments, and on land, the growing use of brass surveying instruments were met by highly skilled craftsmen concentrated in the port cities.[82] Some of them acquired widespread reputations and public recognition for the quality of their work. In Philadelphia, as Brooke Hindle has shown, Benjamin Franklin led a movement to create an American science patterned after the practice of science on the Continent. One of the outstanding colonial contributors to science and a member of the Royal Society of London was the clockmaker and mathematical instrument maker, David Rittenhouse, who made his own telescopes. Later treasurer of Pennsylvania (1777–89) and director of the U.S. Mint (1792–95), Rittenhouse is credited with introducing the use of spider lines in the focus of an instrument. His extremely accurate orrery, or mechanical planetarium, constructed using copper plates and brass gears, gained him an award of £300 from the Pennsylvania Assembly in recognition of his skill and mathematical genius.[83] Rittenhouse and Paul Revere, with their unique contributions to American society, became the most prominent representatives of that important class of colonial craftsmen, the coppersmiths.

Far simpler in origin than the metal crafts already noted and far more limited in practice during the colonial period were the practices of pewterer and tinsmith. Pewter was a metal often found on the list of colonial household items. First wood, then pewter, and, finally, in the early nineteenth century, china were the materials of fashion for plates and dishes; in the colonial period the use of pewter was at its height. All but the poorest families owned at least one or two pewter items, and wealthier families accumulated substantial inventories of pewterware, including porringers, tankards, coffeepots, and candlesticks. In the eighteenth century, pewter plates were in vogue in England and the colonies, although the sturdy wooden trencher remained the standard eating item in the majority of settlers' homes.[84]

Pewter is an alloy of tin, containing either copper, antimony, lead, bismuth, or some combination thereof, depending on the ultimate purpose, desired finish, and cost. Since tin melts at a relatively low temperature, simple pewter objects such as plates, spoons, and porringers are formed by making a mold in some suitable material. Sand or even wood is sufficient for single castings, but gun metal (bronze) is used when many items of the same shape are to be made. The molten metal is poured into the heated mold and the solidified casting is trimmed and polished. Plates and candlesticks are hammered or turned on a lathe to produce a uniform appearance. Intricate items such as coffeepots require the casting of several separate pieces which subsequently are joined by soldering.[85]

The techniques for casting and working pewter were relatively simple and could be performed in the colonies, but the first colonial "pewterers" probably were engaged in the repair and refinish of imported ware, such as the rewelding of broken spoon handles or the resoldering of tankards and other composite pieces.[86] Because of the relative softness of pewter, it did not stand up well to rough handling or repeated usage. More extensive work, such as the casting of large pewter plate, required the use of massive bronze molds which had to be imported. Although these molds were expensive, most pewterers owned a few. To cast pewter, however, the smith needed a source of raw materials. Since tin and antimony (bismuth was little used in the eighteenth century) were not found in America during the colonial period, the manufacture of pewter depended almost exclusively on old or broken pewter as a raw material. Its resale value was about 50 percent of the price for new pewterware.[87]

The largest amount of pewter used in the colonies came directly from England. Among manufactured goods, pewterware ranked at or near the top in value of English exports for most years throughout the eighteenth century. In 1760 alone, the value of pewter shipped to the American colonies exceeded £38,000, greater than the total value for silver plate shipped during the decade from 1760 to 1770.[88]

The number of pewterers in the colonial period was relatively small; not more than a dozen have been confirmed before 1750. Possibly not more than nine were active in 1750 in New York, Philadelphia, and Boston. The latter city, with five, was the center of the trade in the colonies. The pewter industry subsequently evolved in or near the major port cities, which were the centers of population—Philadelphia, New York, New Haven, Newport, Providence, and Boston—or in towns readily supplied by water from the main ports of entry, such as Hartford and Middletown, Connecticut, and Taunton, Massachusetts. Although temporarily disrupted by the Revolution, the trade grew and spread to more inland cities until, by 1825, over eighty skilled pewterers were practicing, with New York and Philadelphia ahead of Boston in the numbers involved.[89]

Although the first pewterers in America had learned their trade in England, the majority of later craftsmen were trained in the colonies. For example, one of the first Newport pewterers, Lawrence Langworthy, immigrated directly from England. The first Providence pewterers, Samuel Hamlin and Gershom Jones, arrived in Rhode Island from Middletown, Connecticut, where they had been taught their trade by the English-trained Danforth family of the Connecticut River Valley.[90]

Tinsmithing, involving the same basic material as pewter, was a craft that developed late in the colonial period. Up to 1740, what little tinware there was had been imported from England. Edward Pattison, a Scottish immigrant, was the first to make tinware in the colonies, shortly after he settled in Berlin, Hartford County, Connecticut, about 1750. Because of their brightness and ease of cleaning, his wares proved so popular that Pattison began to teach the trade to his neighbors so he could keep up with the demand. The Berlin area quickly became the center of tinware manufacture in the colonies. To market their wares, they dispatched wagons to other towns. Stopping at farms along the way, they developed a steady trade in tinware and later added wooden clocks and other household items. This method of distribution subsequently gave rise to the ubiquitous Yankee peddlers.[91] Production of domestic tinware was suspended during the Revolution because of a shortage of tin. Existing supplies were, for the most part, used to make lightweight cartouche cases for the Continental army. After the war, production resumed on an increased scale. Before the end of the century, many Connecticut tinsmiths engaged in the design of machinery to shape the easily formed tin products. Devices for rolling, cutting, bending, and making seams were made for sale to tinsmiths. Pattison's craftsmanship had led to the development of one of the first machine-making industries in the United States.[92]

The goods and services provided by the metal craftsman were of importance to all elements of colonial society, but a singular feature of the practice of the metal crafts was their disproportionate concentration in the northern colonies. Some of the factors accounting for the phenomenon have been noted in the discussion of the individual crafts.

Because the concentration of skilled metalworkers begun in the colonial period had far-reaching consequences for the changing patterns of commerce and in the growth of infant industries presaging the subsequent industrial development of the nation, a more comprehensive look at its causes is warranted.

Although the towns and cities of eighteenth-century America contained a definite minority of the total population, they did contain substantial wealth, a consequence of the thriving maritime commerce. First and foremost, this wealth drew the craftsman to the towns. With the possible exception of the blacksmith, whose craft was universally employable, the specialized skills of the metal artisan catered to demand generated by a degree of affluence and the more complex requirements of urban society. The craftsman, in turn, was accorded a middle-class economic status in urban society because of the relative importance of his work. In his definitive study of social classes in the colonial period, Jackson Turner Main found that artisans or skilled workers in general occupied a position intermediate in the economic scale, between farmers, professional men, and tradesmen on the one hand and mariners, soldiers, and fishermen on the other. Within the artisan class, however, the metal craftsman was at or near the top. The majority of artisans owned homes, and a few, notably blacksmiths and silversmiths, achieved a comfortable degree of prosperity. In terms of social prestige, though still considered as part of the ranks of "workmen," the relative fluidity of colonial society, particularly in the late eighteenth century, did permit some craftsmen to rise to the rank of "gentleman," as we have noted in the case of the Syng and Richardson families, silversmiths in Philadelphia. The social potential inherent in urban living was an additional inducement for metal artisans to concentrate in the northern colonies.[93]

The presence of the metal trades in the towns also stemmed from more general economic problems of supply and demand. The copper-, silver-, tin-, and pewtersmiths depended upon imports for most of their raw material. The balance came from scrap, and this also was easier to obtain where the concentration of people and money to buy goods was highest. When it came to selling his wares, the smith again found easier access both to markets in the populous seaboard towns and to the major transportation network, the waterways extending north and south along the coast and the rivers leading to the interior. Overland transport was slow and expensive.[94] As a corollary, it is not surprising to find that the blacksmith and wheelwright were the craftsmen most common in the rural areas, since their products were heavy and relatively inexpensive and could least bear the high freight charges for overland transport. The same factors also tended to influence the distribution of craftsmen throughout the colonies, the more densely settled North again being more attractive than the sprawling, rural-agrarian markets of the South, even though slightly over half the population of the colonies lived below the Chesapeake in 1775.

The concentration of craftsmen in the North, particularly trained silver-, copper-, and pewtersmiths, had important consequences for the subsequent development of the United States. Their presence there and their relative absence in the South fostered the growth of manufacture in the North, particularly after the Revolution interrupted trade with Britain, making the United States more dependent on their own resources. When industrialization began on an increasing scale in the early nineteenth century, the skills in metalworking already prevalent in the North practically ensured its fullest development there. The iron industry had centered in Pennsylvania before the Revolution on the basis of natural resources. Its center remained in Pennsylvania, shifting westward in the early nineteenth century. But many of the industries that developed in the nineteenth century in the northern states—the manufacture of silver, copper, brass, clocks and watches, machinery and machine tools—trace their origins to the presence of the colonial metalworker.

❦ 6 ❦

Colonial Iron: Regulation and Rebellion

As to what regards Iron Manufacture, let them [Americans] be restrained from making Anchors for Shipping, which weigh above a Hundred and Twelve Pounds; from making all Manner of Ordnance, or Great Guns, or even Shot for such; and from making Fusees, Muskets or Pistols; in short, all sorts of Fire-Arms and Swords, as also all Manner of Locks, and Joint Hinges, with all Manner of Tools for Artificiers; as Carpenters, Sawyers, Joyners, Bricklayers, and all Sadlery and Cutlery Iron-work, together with all sorts of Nails, less than Weight-Nails, Horse-Nails only excepted. But let them not be deprived of mending any of the aforesaid Utensils.—Anon., British, ca. 1745[1]

Should the Government at any Time think fit to take these Gentlemen's advice, I shall not take it upon me to say, what the Consequences might be.—Archibald Kennedy, American, 1750[2]

The growth of the metal crafts in the eighteenth century, particularly the allied fields of blacksmithing, closely paralleled the increasing availability of iron from domestic furnaces. Even as England struggled to assure herself of an adequate supply of iron for home use, her American colonies, hitherto almost completely dependent on her for all classes of manufactured goods, were beginning to achieve a degree of self-sufficiency in the important area of metals manufacture. The significance of the fact that total dependence on England for manufactures was neither satisfactory nor adequate had been recognized as early as 1644. In that year the General Court of the Massachusetts Bay Colony broke the monopoly given to the Company of Undertakers when Hammersmith failed to sustain an adequate supply of iron to meet the colony's needs.[3] During the remainder of the seventeenth century, however, the desire to have domestic manufactures did not make up for the lack of finances and technical ability. By the early eighteenth century, both were available in America, and changing political and economic conditions only reinforced the need to institute domestic manufacture in such basic commodities as iron.

British mercantile theory still had as a fundamental tenet the maintenance of colonial economic subservience to the mother country, but the legislative maneuverings over the regulation of iron manufacture before 1720 hardly inspired acquiescence on the part of the colonies to such a policy.[4] Instead, the American response was to develop the resources to meet the need for iron through colonial manufactures. Some feared that the ultimate consequence of that colonial response, if not checked, might well be nothing less than economic independence. Furthermore, the realization that the creation of economic independence might one day lead to thoughts of political independence was appreciated by at least some of the participants in the colonial scene and in England. Sir John Trenchard, writing as Cato amid the debates over the regulation of colonial iron manufacture, clearly saw the danger. The northern colonies, with growing populations and possessed of abundant supplies of "Timber, Hemp, Iron and other Metals," must be carefully regulated or they would, "by the natural Course of humane Affairs, interfere with most Branches of our Trade, work up our best Manufactures, and at last grow too powerful and unruly to be governed for our Interest only."[5] After 1740, Trenchard's warning took on greater significance because of a new direction clearly discernible in the American iron industry. That direction, to serve the needs of the American market, ran directly counter to British iron interests, which stepped up their clamor for regulatory legislation. The rapid growth of a colonial iron industry, therefore, demanded that attention be given to its regulation, since iron was one obvious key to economic independence.

The British government finally felt compelled to take steps to limit further expansion of the manufacture of iron. The colonists, however, found regulation the more onerous as the industry came closer to supplying fully their needs for iron. The clash of interests, British attempts at regulation versus the continued defiant development of an American iron industry, became the central theme in the history of iron manufacture during the balance of the colonial period. Colonial governments defied Parliament by continuing to encourage new facilities and by failing to enforce regulatory legislation. As the spirit of political independence emerged, one of the issues seized upon was the attempted regulation of the iron industry. By 1780, iron was being made in all of the thirteen colonies except Georgia. The existence of a viable iron industry, which had grown to equal that of Great Britain at the time of the Revolution, was an important factor in the victorious struggle of the colonies to achieve independence.

Aside from the efforts of a few entrepreneurs such as Colonel Spotswood, who was an official agent of the British government in Virginia, or the British stockholders of the Principio Company in Maryland, the initiative for building colonial forges and furnaces in the eighteenth century came from private citizens in the colonies. Government support and encouragement, reflecting colonial determination to establish domestic manufactures, were slower to develop. As early as

1719, Maryland passed an act withdrawing the previously held rights of the proprietor to royalties on minerals and authorizing a land grant of one hundred acres and freedom from levies on workmen for four years to any who would set up forges in the colony. The act coincided with the activities of Onion and Russell, who were establishing the Principio ironworks at the head of Chesapeake Bay, and may have been related to their efforts.[6]

Local need provided the inducement for the series of monopoly rights issued by the Connecticut General Court to petitioners proposing to produce or process iron within that colony. In 1722, with parliamentary debates that reflected a strong sentiment against the erection of iron manufacture in the colonies fresh in their memories, the court granted Ebenezer Fitch a fifteen-year monopoly to set up a slitting mill to make nails.[7] Connecticut possessed no operating ironworks at the time. Shortly thereafter, when Samuel Higley of Simsbury petitioned for a monopoly to "transmute common iron into good steel sufficient for any use" and claimed to be "the very first that ever performed such an operation in America," the court granted him exclusive rights to make steel for ten years.[8] Ignoring continued remonstrances to Parliament from British iron manufacturers, the court issued a similar fifteen-year monopoly to Thomas Fitch, George Wyllys, and Robert Walker, Jr., barely ten years later. Whereas Higley could not sustain his production, observing "that all due encouragment be given to promote such profitable and useful manufactures in the Colony," Fitch and company made good their claim by producing at least one-half ton of steel to the satisfaction of the court.[9]

Rhode Island, another colony with no operating ironworks, waited until the 1730s before the government took a hand in encouraging iron manufacture. Early in the decade, the General Assembly approved a loan of £200 to Samuel Bissell of Newport to make nails for the colony.[10] In 1741, they authorized the construction of a permanent dam across the Pawtuxet River for a proposed ironworks, in contravention of a 1735 law that forbade the erection of permanent obstructions so that the streams might be open for fish runs. That action established the priority of ironworks over other interests in the colony, a priority maintained throughout the balance of the colonial period. When former Governor Stephen Hopkins and Moses Brown asked for a similar exception to erect a furnace in 1769, noting that "a fishery . . . will not, in any measure, equal the advantages which must be derived to the community, by carrying on so large and so useful a manufacture," the assembly granted the petition.[11]

In Pennsylvania, no special inducements by government were necessary to encourage the steady growth of the iron industry, financed by private initiative. The colony, however, stood ready to support that growth if requested, as in 1737, when the General Assembly responded to a petition by ordering the building of roads to connect the Coventry Ironworks and Redding's Furnace "to the most convenient established

Road leading to Philadelphia."[12] The following year, New York, one of the colonies most staunchly loyal to the interests of the British crown but still with no ironworks, passed "An Act emitting Bills of Credit for the payment of debts and for the better support of the Governor of this Province and other purposes therein mentioned." Lieutenant Governor George Clarke reported to the Lords of Trade that one of the reasons that moved the assembly to pass the act was the desire to retain specie in the colony to be used for the development of manufactures, the assembly "intending if they could to enable the proprietors to build furnaces and forges for pig and barr Iron."[13]

Much activity in the colonial legislatures during the latter half of the decade of the 1730s represented a reaction to a depression in the British iron industry that once more adversely affected the supply of pig and bar iron. In England one initial effect of the depression was a clamor from British fabricators to increase the production of pig iron in America for export. The full consequences of the depression will be treated when considering the regulatory debates in Parliament. Here it is important to note that, because the depression diminished the supply of manufactured goods as well as basic iron, the colonies endeavored to erect slitting mills and steelworks, manufacturing facilities, as well as blast furnaces for making pig iron. That course of action was not in line with British thinking on colonial iron.

The history of British efforts to regulate the manufacture of iron in the colonies coincided, as has been noted previously, with the rise of an indigenous industry after 1715. During the seventeenth century, the British government had no policy regarding mining, smelting, or processing metals except for the perpetual reservation to the crown of two-fifths of all gold and silver. Development was left to the initiative of private enterprise, subject to any restrictions, such as enumeration or import levies, which Parliament guided by the crown and colonial officials might see fit to enact. During the seventeenth century, in the case of Hammersmith, and the early eighteenth century, as the creation of the Principio Company demonstrated, there was a willingness to promote the smelting of iron in the colonies and to invest in the erection of furnaces. The impetus came from English merchants and ironmasters intent on supplying the mother country with pig and sow iron for reprocessing into finished goods. The colonists, however, began to perceive the benefits that could be realized from manufacturing ironware on their side of the Atlantic. But efforts to do so would conflict with the interests of the second group of British businessmen, those engaged in producing the finished goods for sale in England or export abroad. Such was the state of affairs in 1715 when the British iron fabricators persuaded Parliament to pass legislation that raised the price of bar iron in the colonies.[14]

Following the royal proclamation in 1717, which terminated all trade with Sweden, including the importation of high-quality iron, a series of petitions were introduced into Parliament seeking to encourage the production of pig and bar iron in the colonies and to remove import duties on such material. Memorials were received from the colonies also, and men such as William Byrd appeared before the Board of Trade to argue for aid in establishing ironworks in Virginia.[15] At the same time, however, members of the lobby representing the iron-mongers, smiths, traders, and manufacturers united in efforts to pass legislation that would have expressly prohibited the manufacture of any finished goods in the colonies. Their lobby proved initially to be more effective than the one representing colonial interests. The Naval Stores Bill of 1719 contained the clause: "That from and after the 25th day of December, 1719, no Forge going by Water, or any other Work whatsoever, shall be erected or kept up in any of the said British Plantations, for the making, working, or converting any Sows, Pigs or cast Iron, into Bar or Rod Iron."[16] Although the issue was not then resolved because the bill failed the final passage, the manufacture of iron in the colonies continued to be a subject of debate in the halls of Parliament. It was in the face of such obvious concern for and opposition to manufacturing in the colonies that the Americans turned to the production of finished goods for the local markets. Their actions only added fuel to burning issues across the Atlantic.

In England, the two factions that had formed during the Swedish crisis continued to contend in Parliament for the protection of their vested interests. The manufacturers of iron products favored the encouragement of a greater output of pig and bar iron in the colonies but demanded restriction of the exports from the colonies to England only, ensuring a steady and relatively cheap source of supply for their operations. Furnace and forge owners, however, strove to ban the production of any iron in the colonies. Toward that end, the latter group managed to introduce a clause into a bill in 1729 that would have required the destruction of all forges in the colonies and a prohibition against the building of any new ones. Colonial agents, influential Englishmen who were paid to lobby on the colonies' behalf with Parliament, where the colonies were otherwise unrepresented, fought to have the clause deleted. Joined by British hardware manufacturers who wanted bar iron from the colonies, they finally succeeded.[17] It was becoming increasingly clear to Americans, however, that whatever policy the British government adopted toward the manufacture of iron in the colonies, and the adoption of some restrictive policy almost seemed inevitable, it would be dictated largely by British iron interests. British intentions obviously were running counter to the actions already taken in the colonies.

In 1735, while Parliament continued to grapple with the problem of the growing production of iron in the colonies, a depression hit the British iron industry. Diplomatic relations with Sweden were deteriorating, and high duties had been placed on British goods by the king of Sweden. The problem of obtaining supplies of high-quality iron from that country once more became acute.[18] Parliamentary debates concerning iron manufacture intensified in the midst of the controversy. Pamphlets and newspaper articles estimated that annual iron imports for England amounted to twenty thousand tons, fifteen thousand tons from Sweden alone, at a cost of £150,000, while exports of wrought iron, mainly to the colonies, accounted for thirty-five hundred tons per annum. In the face of potential shortages of iron for manufactures, factionalism resurfaced stronger than before.[19] The tenuous relations with Sweden led to a new movement to encourage the production of pig iron in the colonies. It was proposed that all duties on its importation into England be removed as an inducement to increase further the annual shipments of over two thousand tons of pig iron already being received from the colonies.

British ironmasters and furnace owners, however, again led a countermovement to restrict any iron manufactures in the colonies. For three years, as the depression deepened, the issue was argued in Parliament and in newspapers. Gradually, the forces favoring protection of domestic industry in Great Britain gained the upper hand. In 1738, they forced debate on a series of resolutions in the House of Commons. At the heart of the issue were three propositions that framed the case for protection:

> *Resolved,* That it is the Opinion of this Committee, That the making of Bar Iron in *Great Britain* is greatly decreased.
> *Resolved,* That it is the Opinion of this Committee, That the Manufacturers from Bar Iron in *Great Britain* are greatly decreased.
> *Resolved,* That it is the opinion of this Committee, That the Making and Manufacturing of Bar Iron in the Plantations is one great cause of said decrease.[20]

Further resolutions proposed the restraint of existing facilities and the prohibition of new ones, and, following the debate, a bill was drawn up to that effect. Once again, colonial agents combined with British iron manufacturers to lobby against its passage. It failed on the first reading. With the outbreak of war in 1739, British industry revived, and pressures to control colonial iron manufactures once again ebbed. Another threat to the colonial iron industry had been narrowly averted, but the interests favoring its further growth clearly were on the defensive.[21]

The debates in Parliament over colonial iron manufactures and their effects on British iron interests continued with reduced intensity for the next decade. During that period, the industry continued to expand in the colonies. New furnaces and forges were built in Pennsylvania, Maryland, and Massachusetts, and manufacture commenced in colonies, where heretofore there had been little or no activity. In December 1741, Governor Lewis Morris of New Jersey forwarded a petition from the Council and General Assembly relating to the encouragement of the making of iron in that colony. Governor Morris had inherited the lands and mining rights of his uncle, Colonel Lewis Morris of Barbados, who had pioneered the manufacture of iron in the Jerseys by building a furnace at Shrewsbury in the 1670s. The mining district west of the Watchung Mountains, settled by iron prospectors early in the eighteenth century, had been incorporated as Morris County in 1739 in honor of the governor.[22] With such a background, it is not surprising that Morris warmly endorsed the petition, noting: "I know there are great quantities of Iron mine, and very good both in ye Province of New Jersie, New Yorke & Pensilvania, & I believe sufficient not only to supply Great Britain & Ireland but great part of Europe."[23] Fully cognizant of the pressures being exerted on Parliament to discourage such actions, Morris added that the importation of British goods in payment for American iron would remove the "necessity, and consequently the temptation" for the colonists to undertake the manufacture of such goods on their own, thereby siding with the views held by British ironmasters.

Over the next decade, new ironworks were constructed in all three colonies, including the Warwick Furnace in Pennsylvania, one of the largest in the colonies. The initiative was taken by colonists without any encouragement from the crown or Parliament. A few years after Morris noted the presence of iron in New York, Philip Livingston, member of a politically prominent family, erected a furnace and forge on Ancram Creek in Columbia County. It was an unusual location, about fourteen miles from the water transportation afforded by the Hudson and an equal distance from his mines in Salisbury, Connecticut. In 1744, an advertisement in the New York Weekly Journal offered "Choice Pigg Iron" for sale at £8 per ton "Ready Money, or Rum, Sugar mollasses at Market Price," noting that the iron was made at Livingston Manor.[24] The rich deposits of the Salisbury-Lakesville region later furnished ore for a number of furnaces in the Connecticut Colony.

In 1752, ignoring belated attempts by Parliament to prevent further development of the colonial iron industry, Timothy Ward and a silent partner named Colton built a furnace in the Sterling tract in Orange County, New York, for the manufacture of anchors. This was the beginning of the Sterling ironworks, a major operation by the time of the Revolution, and the forerunner of the extensive iron industry in the Ramapos region of New York and New Jersey that developed in the next few decades. The Ancram ironworks, the first in the colony, and the Sterling works were the largest and most important established in New York before the Revolution.[25]

While new furnaces were being built in America, events in Europe were again affecting the fortunes of the British iron industry. In 1748, the Treaty of Aix-la-Chapelle, which ended the War of the Austrian Succession, once more aggravated diplomatic relations between England and Sweden. Russia and Sweden threatened war against each other. France, England's age-old political foe, was pledged to the side of Sweden, England's industrial adversary, while England was bound by treaty to uphold Russia's interests. In this volatile political context, iron from the colonies loomed as an attractive alternative to the worsening problems of Swedish supply. The government appointed a committee headed by Charles Townshend to draft a bill admitting pig and bar iron from the colonies duty free. A great hue and cry went up from the manufacturing interests throughout England. Petitions once more flooded Parliament defending or attacking such action. The old arguments were paraded by both sides, but the opponents went to new extremes, evoking apocalyptic visions of the effects of American iron on the British manufactures. Not only would the primary iron industry be all but ruined, they claimed, but, since far fewer trees would be cut to make charcoal, the owners of forests would suffer. Even the tanning trade would be destroyed because of a lack of oak bark obtained from the woodcutting.[26] The exact size and extent of the colonial iron industry was as yet incompletely known and understood in England, but its significance had assumed enormous proportions to influential segments of the English people, if even a fraction of the extravagant claims are to be granted any credence. A picture of widespread unemployment, suffering, and economic disruption was drawn by the anonymous writer of the tract, "The State of the Trade and Manufactory of Iron in Great Britain Considered" (1750). There were, he noted, existing facilities in England to produce at least eighteen thousand tons of bar iron per year, worth more than £300,000 and employing twenty-seven thousand hands. American iron, inferior, he claimed, to either Swedish or Russian for the making of steel, could not compete with the foreign iron but would compete directly with British manufactures. The proposed scheme should be defeated or, "Instead of rendering *Great Britain* independent of her Northern Neighbours, it will render *America* independent of Great Britain."[27]

Petitions from the various interest groups were presented to Parliament, and the ensuing debates were long and heated. As Bining observes, "While the finely worded petitions were in accord with true mercantile principles, they really reveal the self-interest of the various parties."[28] The result of conflicting pressures was the formulation of a compromise bill, signed into law by the king in April 1750 as 23 George II, c. 29, and entitled, "An act to encourage the Importation of Pig and Bar Iron from His Majesty's Colonies in America; and to prevent the Erection of any Mill or other Engine for slitting or Rolling of Iron; or Plating Forge to work with a Tilt Hammer; or any Furnace for making Steel in any of the said Colonies."[29]

The law had two distinct parts, each representative of one of the conflicting sides of the issue. The first part, as the name indicates, was designed to encourage the importation of pig and bar iron into England by the removal of the import duty on those products of 3s. 9.5d. per ton. That the economic inducement was so small (compared to the price of iron, £6 to £7 for pig and £11 to £12 for bar) as to have little probable effect on imports seems not to have been considered. The

Nail Rod Rolling and Shearing
This diorama shows a water-powered rolling mill. A number of such works, proscribed by the Iron Act of 1750, were erected in the colonies prior to the Revolution. Slitting rolls or, more commonly, shears such as those shown in the foreground, cut rolled plate into thin strip or nail rod, the primary product of such a facility. (*Courtesy, The Hagley Museum*)

exemption from duty was limited exclusively to the port of London, possibly because that city was removed from the main centers of British production and tended to draw more heavily on supplies of Swedish iron. That provision created resentment among other major port cities, such as Bristol, and when further troubles with Sweden developed during the Seven Years' War, another wave of petitions persuaded Parliament to abolish the exclusive privilege granted to London.[30]

The more significant portion of the act for the colonists was contained in the later passages, beginning with Section IX, stating that "no mill or other engine for slitting or rolling of iron, or any plateing-forge to work with a tilt hammer, or any furnace for making steel" could be erected in the colonies after June 24, 1750.[31] A penalty of £200 was imposed for violation, and the works were to be destroyed. Although manufacturers had endeavored to have all such works banned, works already in operation before that date were not affected.[32]

The immediate effects of the Iron Act were mixed. Americans readily accepted the reduction of the tariffs on their iron, and the amount of bar iron shipped to Great Britain from the colonies rose slightly to an annual average of only 270 tons in the years immediately following 1752. The exportation of pig and sow iron, primarily from the Chesapeake Bay furnaces of the Principio and Baltimore companies, already nearly twenty-five hundred tons per year, remained stable, but exports from New York and Pennsylvania increased from two hundred to one thousand tons as a result of the building of new furnaces.[33] Bar iron exports lagged far behind, New York becoming the leading export colony in 1768. British imports of American pig and bar iron began to rise in the period 1765 to 1770, reaching a maximum in 1771 but dropping off sharply thereafter.[34] The higher imports were not due to the success of British regulation, however, but to a mild recession in the colonies, the result of overexpansion that forced the colonial ironmasters to export larger than usual quantities. By 1771, the growth of the colonial demands had caught up with the new supply and exports dropped. These results were far below the expectations of the bill's authors and of the London merchants who had hoped to benefit from their monopoly position.

The primary reason for the failure to increase the supply of iron from the colonies was that the worst fears of English ironmasters already were being realized. By the middle of the century, the output of the colonial industry was being utilized by the expanding domestic market in the form of utensils, tools, and implements. Illustrative of the reason for the failure of the colonies to increase iron exports was Connecticut's reply to the "Queries" of 1756, particularly the one concerning the presence of mines. While it was noted that iron had been found and developed in sundry places, the answer concluded with the plaint that there was "not sufficient supply for our inhabitants."[35] Connecticut, like most colonies, needed more iron than it could produce. Only in the Chesapeake area was the primary production being sent abroad. Elsewhere, as in New York, exports represented a surplus, all but a small part of the domestic pig and bar iron having been sold in the region to supply local needs. Except for the depression years of 1770–71, Pennsylvania's large production was absorbed almost entirely by the domestic markets.[36]

Before the passage of the Iron Act, the "Queries" sent by the British Board of Trade to each colony at intervals asking, "What sort of manufactures do you receive?" almost invariably had elicited a response including nails, farm tools, firearms, and cutlery. These were among the items the colonial ironworks were attempting to produce.

To ascertain the status of the iron industry in the colonies at the start of the period of regulation, the commissioners for Trade and Plantations specifically requested data on the number of restricted facilities built or operating before June 24, 1750. This time the returns reflected a sharp change in attitude on the part of the colonists toward cooperation with any attempts at British regulation. Some returns were detailed. Connecticut reported one steel furnace, built in 1744, and eight forges with hammers, giving locations and dates of construction.[37] Governor Jonathan Belcher of New Jersey, however, listed one steel furnace and only one slitting mill and one plating mill.[38] New York reported but one plating mill in addition to Livingston's Ancram works, possibly because the Sterling works were in a region still being disputed between New York and New Jersey.[39] In any event, neither state mentioned the Sterling ironworks. Massachusetts reported two slitting mills and Pennsylvania, one. Despite the surprising number of plating mills reported by Connecticut, Pennsylvania and Maryland, with far more extensive facilities, reported only one each, while Massachusetts listed two. Complicating the problem of classifying ironworks was the fact that any forge equipped with a tilt hammer to make bar iron could be converted to a plating mill simply by changing to a hammer head of a different shape. Although reporting only two plating mills, Massachusetts had more than forty forges, many of them suitable, if not already equipped, to make plate.[40] The other governors insisted that none of the specified works existed in their colonies.[41] The figures that were transmitted to the Board of Trade represented a substantial increase in the size of the colonial industry, but, as Bining has shown in the case of Pennsylvania, the returns were far from inclusive, a number of proscribed facilities not being reported.[42]

The provisions for enforcement of the law showed how little grasp the members of Parliament had of the actual state of affairs in the colonies. The colonial governors were to be held responsible to see that no new works were erected on the penalty that they were to be fined £500 for each new works discovered and enjoined from holding office, a far stiffer censure than the one confronting the potential builder. Moreover, beginning with John Winthrop, Jr., who had a long career as governor of the Connecticut Colony, colonial officials had been among the leaders in promoting the development of the iron industry, often having considerable investments in mines, furnaces, and forges. Hence in many cases the initial reaction to the restrictions of the Iron Act of 1750 was one of caution and evasion, the filing of ambiguous and inaccurate reports by the colonial officials charged with carrying out the provisions of the bill.

The American disregard of British efforts toward regulation, bordering on active collusion on the part of the colonial governments, grew stronger with few attempts being made to prevent the erection of new mills and furnaces as proscribed by the Act of 1750. Production of bar iron continued to expand from existing or new facilities directed toward the manufacture of hardware in the colonies, one of the issues that had precipitated the passage of the Iron Act. In New York, by 1760, Philip Livingston was advertising for three good refiners to make bar iron at a new forge, claiming that "Stock will never be wanting,"[43] and two years later Livingston wrote to assure Roger Wolcott of Connecticut that "I should be able to Supply all your forges as fast as they can work it up into bar Iron."[44] Examples of defiance of the Iron Act may be drawn from the records of practically every colony. Illegal slitting mills were established in New Jersey, Massachusetts, and Connecticut.[45] New steel furnaces were openly advertised in New Jersey, including one as early as 1767 in Perth Amboy. Yet, the following year Governor William Franklin reported to Secretary Hillsborough that the iron industry of New Jersey consisted of eight blast furnaces, slitting mills, and plating mills, all of which, he added, "were erected before the Act of Parliament respecting these works."[46]

Because one of the primary sources for identifying ironworks was the governors' reports and these became less specific after 1750, the exact number and type of forbidden works erected between 1750 and the Revolution is not known. What evidence is available, however, strongly indicates that the manufacture of steel for tools and the production of nail stock, which was made by slitting mills, both continued to increase. The greater availability could have come about only through production from unreported and illegal installations. Abel Noble and Peter Townsend of New York, for example, advertised in 1775 for men who understood the making of steel "in the German method," which was how the centuries-old process of cementation was known in the colonies, to operate a forge with "six fires" being built in the colony for that express purpose.[47] New York and Philadelphia papers advertised the availability of steel implements from several facilities already operating in New Jersey. Men who knew the nail trade also were in some demand, in at least one instance by persons who openly proposed to enter into the manufacture extensively.[48]

The failure of the colonial governments to submit accurate returns on regulated facilities to British authorities amounted to nullification of the Iron Act in the colonies without the acquiescence of Parliament. Hence that failure could only be construed as an act of defiance on the part of the colonists, possibly presaging nullification of future unfavorable legislation. It is significant that this occurred during the 1750s, nearly a decade before the more concerted resistance to the Stamp Act cited by many historians as the turning point on the road to independence.

Following the close of the French and Indian War in 1763, British attempts to increase the economic burden on the colonies through the infamous Stamp Act, coupled in 1764 with the enumeration of iron, brought smoldering resentment and covert defiance of regulations into the open.[49] In the debate over the repeal of the Stamp Act, William Pitt, "the Great Commoner," took the side of the colonists on that issue but demanded also that "the sovereign authority of this country over the colonies be asserted in as strong terms as can be devised, and be made to extend to every part of legislation whatsoever; that we may bind their trade, confine their manufactures, and exercise every power whatsoever—except that of taking money out of their pockets without their consent."[50] Many colonists, however, were coming to consider the confinement of manufactures as identical with the "taking of money . . . without consent."

John Dickinson, a distinguished Pennsylvania lawyer with investments in iron,[51] became one of the principal spokesmen for colonial opposition to the Stamp Act and all subsequent attempts by Parliament

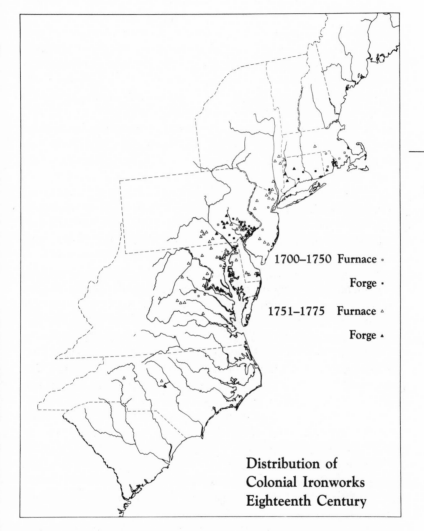

1700–1750 Furnace

Forge ·

1751–1775 Furnace ᐃ

Forge ▲

**Distribution of
Colonial Ironworks
Eighteenth Century**

to tax the Americans. His most famous polemic was the *Letters from a Farmer in Pennsylvania*, written in 1767–68 to protest the Townshend Acts of 1767. Dickinson noted, with pointed reference to Pitt's remarks, "Great Britain has prohibited the manufacturing iron and steel in these colonies, without any objection being made to her *right* of doing it. The *like* right she must have to prohibit any other manufactures among us. Thus she is possessed of an undisputed *precedent* on that point."[52] After quoting Pitt, Dickinson concluded that if Britain's authority to limit manufactures and to levy duties on manufactured goods for the sole purpose of raising revenue were to be conceded, the status of the colonists would be no better than that of "abject slaves."

As the breach between the colonies and the mother country widened, the volume of literature protesting Parliament's ever more stringent efforts to establish its authority increased accordingly. Invariably, reference was made to the regulation of iron manufactures. In 1772, fiery Samuel Adams addressed the Boston Town Meeting with a statement of the colonists' rights. Claiming that the American colonists were "entitled to all the natural essential, inherent & inseparable Rights Liberties and Privileges of Subjects born in Great Britain,"[53] Adams maintained that constraint of manufacture constituted a violation of those rights. Citing such constraint as one of twelve areas of grievance, he declared: "The restraining us from erecting Slitting Mills or manufacturing our iron the natural produce of this Country, Is an infringement of that right with which God and nature have invested us, to make use of our skill and industry in procuring the necessaries and conveniences of life."[54]

The town meeting endorsed Adams's arguments and, with minor additions, had them printed for distribution throughout Massachusetts and to other interested parties. One such party may have been Thomas Jefferson, who drafted a broad and well-reasoned review of the grievances affecting the colonists to serve as a basis for preparing instructions for the Virginia delegates to the Continental Congress. Jefferson noted that by the act of 23 George II, c. 29, "the iron which we make we are forbidden to manufacture, and heavy as that article is, and necessary in every branch of husbandry, . . . we are to pay freight for it to Great Britain, and freight for it back again, for the purpose of supporting not men, but machines, in the island of Great Britain."[55]

While leading spokesmen for the revolutionary cause in the colonies were citing the British attempts to regulate the iron industry as one of their grievances against the crown, the active promotion of the iron industry by colonial legislatures had been growing more flagrant. Connecticut, in 1772, issued an interest-free loan to Aaron Eliott of Killingsworth to purchase bar iron from the colony's forges for conversion into steel in direct contravention of the regulatory laws.[56] Rhode Island, in 1774, granted a petition to hold a lottery for the repair of Coventry Forge, damaged by fire, in which Nathanael Greene, soon to become a general in the revolutionary army, had a major interest.[57]

Although a bill passed by the North Carolina Assembly and Upper House in 1768 "to encourage Iron Manufacture" in the province had been rejected by the governor and council, by 1771 two furnaces were in operation on the Deep River in Orange County, and a third was being built in Rowan County near Salisbury.[58] On May 20, 1775, following receipt of the account of the events of the opening of hostilities at Lexington, people of Mecklenberg County issued a declaration of independence from England. Governor Josiah Martin fled the capitol at New Bern, and in his absence a Provincial Congress was convened at Hillsborough in August. As one of its first acts, the congress passed resolutions authorizing the payment of substantial bounties for the encouragement of manufactures of gunpowder, salt, paper, cloth, and metals. Among the bounties voted was one of £500 for a furnace to produce pig iron and hollowware, £250 for the erection of a rolling and slitting mill capable of making five tons of slit rod for nails within two years, and lesser amounts for the erection of two steel furnaces, the latter installations clearly proscribed by the Iron Act.[59] Inherent in the ultimate act of political defiance by the people of North Carolina was an expression of resentment over British attempts to regulate manufactures. Their defiance of that regulation, even more than the words of Adams or Jefferson, voiced the sentiments of an emerging nation.

Blast furnaces and forges had not been proscribed by the Iron Act. From 1750 to the beginning of the Revolution, new blast furnaces were built at an average rate of over two per year. Pennsylvania and New Jersey witnessed the greatest growth in the production of iron in the late colonial period. In Pennsylvania, a band of furnaces was located in Berks, Lancaster, and York counties to the west of Philadelphia. By 1770, the band stretched westward along the Maryland border into Cumberland County, where Carlisle (1762), Holly (1770), and Pine Grove (1770) furnaces had been built. Additional works continued to be built in the eastern region, perhaps the best known being Mark Bird's Hopewell Furnace (1770) in Berks County, today a National Historic Site.[60]

The pine barrens of southern New Jersey became the site of a series of iron plantations initiated by the redoubtable Charles Read of Burlington, one of the most powerful political figures in the colonies. As a preliminary to building ironworks, Read acquired, by purchase or lease, over ten thousand acres to ensure an adequate wood supply in the plantation pattern. In a feverish burst of activity, Read managed to erect Etna, Taunton, and Batsto furnaces and Atsion Forge between 1765 and 1768.[61] The effort exhausted Read's finances and health, and he was forced to withdraw from the venture, but the ironworks and their small communities persisted well into the nineteenth century.[62] Read seems to have been an unusal type of speculator for that period. His aim apparently was not to make money from iron but from the sale of completed ironworks, ready to produce for the Central Atlantic market. During the same decade, furnaces were erected at Cohansie and Mount Holly (the latter subsequently was destroyed by British troops during the Revolution) and at several places just west of the Palisades.[63] Northern New Jersey also became the location for the complex of facilities known as the American Iron Company, the product of a remarkable entrepreneur, Peter Hasenclever.

Hasenclever, a man of experience in the manufacture of iron and steel on the Continent, arrived in England in 1763 and formed a partnership with two Englishmen, Andrew Seton and Charles Crofts, raising £20,000 sterling for investment purposes.[64] Although investment opportunities were legion in England and Europe, Hasenclever convinced his partners and a number of other reputable persons, including George Jackson, secretary of the Admiralty, that the most glamourous possibilities were available in the manufacture of iron in the American Colonies, including "inexhaustible woods, full of timber fit for coal."[65] One reason for his conclusion was the determination that Great Britain was importing forty thousand tons of bar iron each year, at an average cost of £16 per ton.[66] The following year, he sailed for America and immediately purchased the Ringwood Ironworks Estate in northern New Jersey for £5,000 sterling. It included nearly 50,000 acres of land, 122 horses, 214 oxen, and 51 cows. By 1766, Hasenclever was able to report that he had built or reconditioned one furnace, four forges with eleven fires, one stamping mill, one sawmill, one gristmill, and a large number of auxiliary buildings and was producing iron.[67] The stamping mill was an important contribution to the technology of ironmaking in America. Stamping mills had been widely used in Europe and even Mexico for centuries to crush ores. Hasenclever used it to separate iron from the cinder heaps that surrounded every charcoal furnace, allowing recovery of iron lost in slag removal and greatly increasing the yield of the facility.[68]

In 1765, Hasenclever purchased nearly six thousand acres fifteen miles southwest of Ringwood and began the construction of Charlottenburg Furnace, two forges with eight fires and another stamping mill. The following year, he constructed a dam to create Long Pond to provide a reliable water supply for another ironworks. This third furnace and forge complex began producing in 1768 at the rate of twenty to twenty-five tons per week. In all, Hasenclever built four furnaces (the fourth was at Cortland, New York) and twenty-four forge fires and imported over five hundred German workmen and their families before he was replaced by an expanded and reorganized company in 1767.[69] The German "miners, founders, forgemen, colliers, carpenters, masons, and labourers" imported by Hasenclever with their families remained in the area, their descendants forming a large element in the population of Pompton and West Milford townships well into the nineteenth century.[70]

The failure of the American Iron Company was almost as rapid as its rise. After expending nearly £55,000, Hasenclever was able to achieve only half of his predicted output of five to six hundred tons of bar iron per year. The workers were refractory or sick, he complained, and of fifty-three mines tested only seven produced good metal.[71] Nor was he able to fulfill his intention of erecting a German steel manufactory because, he erroneously noted, such a facility was "not yet existing in the British Dominions."[72] Although an investigation committee appointed by Governor Franklin of New Jersey in 1768 had given a most favorable report of the results of Hasenclever's industry, the company was alarmed at the high rate of expenditure and the low rate of return.[73] This may, in part, have been due to the company's orientation toward British markets or, more likely, as stockholders charged, the failures of the American Iron Company may have been due to Hasenclever's managerial shortcomings. A long series of court suits ensued which eventually vindicated Hasenclever, who returned to Europe in 1769. By that time, however—1793—the company was bankrupt, most of its American holdings having been seized by the colonists in the Revolution and the remainder destroyed or discontinued.[74]

In 1768, before leaving America, a disillusioned Hasenclever had written to Sir William Johnson, "This country is not yet ripe for manufactures. Labor is too high—too much land to be settled. To erect fabrics (factories) is to ruin the landed interest. . . . (T)herefore I think the present zeal to establish manufactures is premature."[75] There were many in the colonies who did not agree. Throughout the colonies the demand for iron was high. Many successful ironworks were being constructed north and south, without exception to serve the American market. Hasenclever's efforts were not premature; rather, the pro-British bias of his company had become anachronistic. The climate of interest that had supported the creation of the Principio and Baltimore ironworks no longer prevailed. These companies, with long-established markets in England, continued to flourish even as Hasenclever's ambitious project foundered. The American Iron Company was the last attempt before the Revolution to develop American iron exploitatively for British interests.

Although the industry expanded most rapidly in the Central Atlantic states, there was also a resurgence of ironmaking in the South, where the industry of the eighteenth century originally had been established to supply material for the British market. Although the southern colonies began producing for their own consumption, they lagged behind the relative development of the northern colonies. Also, the major facilities of the Tidewater continued to produce materials for shipment to England, Maryland and Virginia shipping over one-half of the total amount of pig iron exported by the colonies as late as 1773. At the same time, a coastal trade developed between the northern and southern colonies in which iron and fabricated iron products were prominent. Tidewater Virginia became one of the leading importers of colonial iron, followed by South and North Carolina in that order.[76] Although no detailed breakdowns of imports for Virginia or South

Carolina are available, the records kept by James Iredell for Port Roanoke, North Carolina, show that from July 1771 to mid-1776 over four hundred thousand pounds of iron hollowware, axes, nails, and bar iron entered the colony through that one port from northern colonies.[77]

Two influences combined to shift the mode of production in the newer forges and furnaces to manufactures for local consumption, after the pattern followed in the northern colonies. The first factor was location. As population moved westward away from the Tidewater area, access to direct water communication with England diminished, which greatly increased transportation costs for imported goods. The second factor was that a main line of settlement was established along the route of the great Wagon Road heading southward and westward from its origin at Lancaster, Pennsylvania, close to the major iron-producing area in the colonies. The technology of iron manufacture followed in the wake of the expanding population. In Virginia, iron manufacture was extended into the Piedmont with the construction of Bear Garden Furnace at the site of an earlier forge in Buckingham County about 1765, followed by a number of others in the next decade.[78] The three North Carolina furnaces were in the Piedmont, as were the first ironworks in South Carolina, forges erected about 1773 to utilize the magnetic iron ore in the upper colony.[79]

By 1775, on the eve of the Revolution, more than eighty furnaces had been built in the colonies since 1715, with Pennsylvania leading with twenty, followed by Maryland with seventeen, and Virginia and Massachusetts each with fourteen. New Jersey, New York, North Carolina, Rhode Island, and Connecticut could claim at least one operating furnace each.[80] The average charcoal furnace was capable of producing three hundred tons per year, while a few giants like Reading and Warwick in Pennsylvania yielded at least eight hundred tons.[81]

In addition to the furnaces, the colonists possessed at least 175 forges. These were concentrated in the northern colonies of Pennsylvania, New Jersey, and Massachusetts, primarily for the purpose of converting pig iron from the furnaces to bar iron suitable for the manufacture of tools and implements.[82] The number of furnaces and forges in the colonies exceeded the number of corresponding facilities in England and Wales in 1775. The overall production of the industry from furnaces and bloomery forges probably was on the order of thirty thousand tons per year according to Bining's estimate, approximately one-seventh of the total world production at the time. This was sufficient to rank the colonial iron industry third in the world in output, behind the major sources for pig and bar iron which were Russia and Sweden.[83] More significantly, it meant that the American colonies, with a much smaller population, were outproducing the iron industry of England and Wales in that basic commodity as the break between them neared.

The trail that started in the forests of eastern Pennsylvania early in the eighteenth century had led in just two generations to the creation of a major industry in America, sufficient to supply the needs of a growing population and, subsequently, to wage a war for independence.

7

Metals and the Revolution

The business of casting Cannon and making fire arms is of infinite importance to this Continent and cannot be too much encourag'd. —Colonel Henry Knox, June 1776[1]

If the colonization of North America by the English-speaking peoples provided opportunities for experiments in the creation of utopian societies, the American Revolution saw those peoples embarked on yet another and greater utopian venture—the creation of, as they conceived it, the perfect polity. The earlier events contributed to the demand for metal products in the colonies, which facilitated the introduction and growth of metal technologies. Those technologies in turn aided the formation and helped to sustain a distinctively American society. After a century of slow growth, that society rapidly matured in the decade following the Peace of Paris. The increasing alienation between the colonies and the mother country during the 1760s and early 1770s was, in large measure, a consequence of the recognition by Americans that they no longer were colonial Englishmen but possessed a unique identity, a concept they attempted in vain to explain to an uncomprehending crown and Parliament. The completion of that social revolution stemming from the utopian ideals of the seventeenth century fostered a rise in utopian political ideology in the period preceding the Revolution.[2]

Reflecting on the developments that culminated in the War for Independence, John Adams wrote, "The revolution was complete in the minds of the people, and the Union of the colonies, before the war commenced in the skirmishes of Concord and Lexington on the 19th of April, 1775."[3] Adams, of course, referred to the revolution in political thought, the establishment of the conviction that a new and more perfect society had evolved in America. When the farmers of Massachusetts stood their ground in the face of British regular soldiers and "fired the shot heard 'round the world," they were but giving final expression to what Adams and other Americans already considered a nearly accomplished fact. The ties of empire, of cultural and political dependence, that had bound the American colonies to Great Britain had been broken. Yet, although the cultural and political revolutions may have run their course, another revolution was beginning, one requiring muskets and bullets for the Minutemen, cannon and shot for armies and fleets, and all the other material accoutrements of war. The creation of the new utopia, like the old, had requirements for metals and fabricated metal products.

The individual colonies, long accustomed to dealing with England as independent entities, took the initiative in purchasing arms and ammunition and in encouraging their domestic manufacture. Months before the opening battles at Lexington and Concord, their separate and uncoordinated programs had begun to enlist the efforts of gun-smiths and iron founders to provide weapons for defense. Once the war began, a "national" arms procurement program was pieced together by the Continental Congress. Especially pressing was a need for field artillery to support Washington's forces. The Continental program often conflicted with the provincial efforts, however. Thoughout the war, a shortage of lead for bullets caused concern to the Americans and inspired diverse efforts to obtain the once lightly regarded metal. The utilization of the metal technologies already present in the colonies, the facilities for smelting and casting iron, and the skills of the craftsmen contributed greatly to the successful prosecution of the war and the founding of a new nation.

In the spring of 1774, Parliament, at the instigation of Lord North, debated and passed a series of acts designed to ensure a proper dependence of the colonies upon the crown and Parliament of Great Britain. The new legislation was a response by Parliament to the outrageous rebellion against royal authority by the people of Boston in destroying several hundred cases of tea and discharging their contents into Boston Harbor. Three of the new acts—the Boston Port Act, the Government of Massachusetts Act, and the Administration of Justice Act—were specifically intended to limit and proscribe the "licentious-ness" of the people of Massachusetts. Together with a new, more restrictive Quartering Act, they comprised a body of coercive legisla-tion which the majority of the colonists soon labeled as "intolerable."[4] In a series of resolves passed by a convention at Suffolk in September, the towns of Massachusetts denounced the acts as unconstitutional and not to be obeyed. More ominously, the Suffolk meeting also recom-mended that citizens begin to arm themselves and prepare to defend their rights, by force if necessary. When the "Suffolk Resolves" were ratified by the delegates to the First Continental Congress a few weeks later, the road to confrontation was open.[5]

In England, the Privy Council responded to the new threat by drafting orders to the colonial governors to prohibit the importation of gunpowder or any sort of arms or ammunition to the American colonies. The orders were contained in a circular letter to the governors drafted by Lord Dartmouth.[6] Matters came to a head when the king's orders reached the governor and council of Rhode Island on December 5, 1774. Its reading was followed by an immediate order to Colonel Joseph Nightingale of the militia to remove all but four of the cannon and all powder, shot, and stores from Fort Island at the mouth of Narragansett Bay.[7] Four days later, six twenty-four-pounders, eighteen eighteen-pounders, and fourteen six-pounders were taken to the town of Providence "from wherever," it was reported, "they may easily be conveyed into the country, to meet the Indians and Canadians with which the colonies are threatened."[8] The council then authorized the purchase of four brass field pieces for the militia, in direct defiance of the regulation.[9] Public reports notwithstanding, the council clearly perceived the real source of threats to the colony's well-being. Arriving in the colony one week later, Captain James Wallace of the British fleet investigated the seizure and reported to his superior, Vice-Admiral Samuel Graves: "A procedure so extraordinary caused me to wait on the Governor, to inquire of him, for your information, why such a step had been taken. He very frankly told me, they had done it to prevent their falling into the hands of the King, or any of his servants; and that they mean to make use of them, to defend themselves against any power that shall offer to molest them."[10]

Rhode Island sent a copy of Lord Dartmouth's letter, together with a description of the actions taken, to the Massachusetts Provincial Congress, an extralegal body established to conduct affairs in the province in defiance of the coercive acts. They, in turn, dispatched Paul Revere to New Hampshire with a report of Rhode Island's action. There, the citizens of Portsmouth, led by John Sullivan (later to become a major-general in the colonial army), responded to the news by seizing Fort William and Mary at the entrance to Portsmouth Harbor. The guard fired a single volley of small arms and light cannon before giving way to the mob, who tore down the British flag, the first such act in the colonies. Within two days, over one hundred barrels of powder, a number of muskets, and eighteen small cannon had been removed before the arrival of a British warship forced the insurgents to leave.[11] Months before the battles of Lexington and Concord, British troops had fired on a band of American colonists who were attempting to arm themselves for an impending conflict in defiance of the crown.

The seizure of cannon by Rhode Island and New Hampshire showed an awareness of a major problem confronting the colonies in the imminent event of war. Gunsmiths were present throughout the colonies able to manufacture and repair muskets and rifles. Other craftsmen, like the blacksmith or iron founder, could supply bayonets and the less conspicuous essentials of war, such as shovels, axes, cooking pots, and cartridge tins. In 1775, only cannon could not be readily obtained in the colonies. Numbers of cannon suitable for little more than coastal defense were scattered along the Atlantic seaboard in fortifications protecting the approaches to harbors and rivers. The largest stores of cannon were in and around the major seaports, but these invariably were controlled by British troops. Another body of small cannon, purchased primarily for use on merchant ships to discourage piracy, were privately owned by merchants or ship captains. In the event of war they could be used to arm privateers but hardly were suitable for engaging British naval vessels. The colonies themselves owned not a single cannon. For land warfare, the almost total lack of brass field pieces and carriages would be a particularly acute problem. To support an army, the entire train of field artillery possessed by the British colonies at the start of the Revolution consisted of only four cannon originally belonging to Massachusetts.[12]

In October 1774, the Massachusetts Provincial Congress, acting in accordance with the recommendations of the Suffolk Resolves, had created a committee to consider the ordnance needed by the province. Their report recommended the purchase from private sources or abroad of twenty field pieces, four mortars, tons of shot and shell, and one thousand barrels of powder at a cost of over £10,000, together with five thousand stand of arms and bayonets, with flints, at another £10,000.[13] On November 15, the Committee of Safety authorized the purchase of seven large cannon on the best possible terms, with the admonition to get them out of Boston to a more secure location. Depots for arms and ammunition were established first at Concord and Worcester, later at seven additional towns. In February, the Committee of Safety assigned the first four brass field pieces purchased to Concord, where they were almost captured by the British raid in April.[14]

The efforts of Massachusetts to arm its militia and to place the colony in a state of war-readiness marked the beginning of a significant chain of events. Too often, in considering the American Revolution, we conceive of it as a consolidated effort on the part of the thirteen colonies to resist British tyranny. The historical evidence hardly supports such a concept. Before the Revolution, and even down through the period of Confederation following it, each colony maintained and jealously asserted a separate, independent existence. No central government or common council existed on this side of the Atlantic before the convening of the extralegal body of the Continental Congress. The only common bond shared by Americans, north and south, was their colonial status within the British Empire. Parliamentary laws affecting their trade and customs impacted with more or less severity not only on the thirteen colonies but upon the Canadas and islands of the West Indies as well. Although a sense of injustice against the crown and Parliament was pervasive along the Atlantic seaboard in 1774, the initial reaction was not to establish unity against a common enemy but to take independent action to preserve the well-being of each colony, following long-established tradition. In this, the northern colonies, long the most independent of British possessions, led the way, beginning with Massachusetts's attempts to secure arms for its militia.

Rhode Island followed its act of defiance by authorizing the purchase of sixty heavy cannon from any available sources in January 1775.[15] Following the outbreak of hostilities at Lexington and Concord, other states began to act. In the process, they paved the way for the large effort of arms procurement that contributed greatly to the successful prosecution of the revolutionary war effort.

After the loss of cannon at the forts in Rhode Island and New Hampshire, the Admiralty removed the small battery at Penobscot in what is now Maine before a similar fate befell it.[16] The large military force quartered in Boston precluded any attempts to take the cannon from fortifications in the harbor there, but no such protection was available in Connecticut, where the batteries protecting the approaches to New London were seized in December 1774.[17] In all the northern colonies by the spring of 1775 the only large store of cannon not yet seized by insurgents, other than at Boston, was at the port of New York, perhaps the most pro-British of the northern colonies. Sentiments, however, could not be regarded as sufficient protection for these important cannon, so Vice-Admiral Graves dispatched the armed sloop *Asia* to discourage their removal. It was suggested that they be removed by ship rather than let them fall into colonial hands,[18] but the presence of a warship in the harbor proved to be an inadequate deterrent. In August 1775, a group empowered by the Provincial Congress of New York attempted to remove the twenty-one naval cannon forming the battery at Fort George at night. A harmless broadside from the *Asia* interrupted the effort, but eventually they were successfully moved to one of three gun parks in Westchester where many small cannon, formerly the private property of New York shipowners, also were collected.[19] Subsequently, in early January 1776, the stored ordnance, nearly the total amount possessed by the colony (about three hundred pieces), were spiked and smashed by unknown parties. An investigation pointed toward Tory sympathizers. By June, however, repairs had been made on much of the damaged artillery, and the Continental forces had prepared for the defense of New York by installing 121 cannon and 19 mortars in fourteen batteries.[20]

The capture of British cannon by colonial militia served as a temporary expedient. Although important in giving the colonies a sense of having taken the initiative and of depriving British forces of the use of established provisions, it was evident that the supplies thus obtained were totally inadequate to protect the extensive coastline from naval attack or invasion. If the seizure of British cannon was to be anything more than a symbolic act of defiance, the colonies would have to arm themselves fully to defend the independence they were asserting. There was, at this early point in the conflict, no thought of waging an offensive war which, in any event, was beyond the capabilities of the individual colonies. Even defense required the massive infusion of arms, particularly cannon. Rhode Island, quickly followed by Pennsylvania and Connecticut, was the first to perceive the means to solve the problem of obtaining an adequate supply of cannon, by turning to the iron industry within its borders. In August, Nicholas Power of Warwick and Jacob Greene, brother of revolutionary General Nathanael Greene, were ordered to obtain eight field pieces for the Rhode Island militia, to be cast in the colony. In December of the same year, another request was made for the purchase of cannon.[21] Early in 1776, the proprietors of Hope Furnace submitted an offer to cast cannon for the state at a price of £35 a ton. Eventually, Hope was to cast a substantial amount of cannon and shot, but in the midst of the turmoil that accompanied the start of the Revolution, the council rejected the Hope proposal, suggesting that lighter cannon would be just as serviceable at only £30 a ton.[22]

Connecticut was only a little slower than its neighboring colonies to prepare for defense. In December 1775, the Council of Safety was empowered to find ways to supply cannon and ammunition for the colony. Although the great majority of colonial ironmasters were sympathetic to the revolutionary cause, Richard Smith, who owned a forge and furnace at Salisbury, was not. At the start of the hostilities, Smith fled to the protection of the British at Boston, and in January 1776, Connecticut seized his furnace for a cannon manufactory. Lemuel Bryant of Middleborough was hired as cannon founder, and by June orders were placed for ten tons of assorted shot. In July it was reported that eighteen nine-pounders and other cannon already had been cast, and before the month was over the operation was proceeding so smoothly that Connecticut confidently voted the loan of twenty cannon to New York for its defense.[23] Surviving records are not sufficient to estimate the output of the Salisbury furnace, but some indication of its scope may be gathered from the legislative session of November 4, 1776. On that one date, it was voted to release twenty-five tons of pig iron cast at the furnace for conversion into steel and to distribute seventy-two cannon and mortars of assorted sizes for coast defense and for outfitting ships, together with the necessary shot and swivels. Yet Salisbury was not one of the larger iron furnaces. A request to cast thirty-pounders for outfitting Continental frigates had to be rejected because the furnace lacked the capacity for such large castings. It continued to produce iron both for the state and the Continental army until 1783, but activity slackened after the winter of 1777–78. By then, General John Burgoyne's defeat at Saratoga had removed the last major invasion threat to New England.[24]

In the Middle Atlantic states, Pennsylvania, with the largest number of ironworks in America, followed the lead of Rhode Island. A cannon committee consisting of leading citizens was appointed in November 1775 to procure cannon "as soon as possible." In particular, David Rittenhouse, renowned for his scientific achievements, rendered outstanding service in the mobilization of the iron industry to the war effort for the colony. In the next year, at least ninety cannon were cast by Reading and Warwick furnaces within the colony, and tons of shot and shell were ordered from Batsto and Atsion furnaces in New Jersey.[25]

The efforts of Pennsylvania to obtain cannon illustrate a problem that plagued the colonial war effort throughout most of the Revolution—the nearly total lack of coordination of programs between the individual provincial congresses and the Continental Congress sitting in Philadelphia. Once the shooting started, one of the first items of attention for the Continental Congress was the procurement of cannon, particularly field artillery, for the support of Washington's armies. A special committee was created for that purpose. The actions taken by the individual colonies, however, belied the union Adams later remembered. They were taken to bolster the defenses within the colonies, often without consideration of the more pressing needs of the active war against British armies. The cannon and shot Pennsylvania purchased from its furnaces and those in New Jersey were for use by Pennsylvania, not the Continental forces. Although the shooting war had begun in April 1775, not until November 1776 did Pennsylvania think to appoint a committee to confer with its congressional counterpart on the expediency of casting cannon for the common cause.[26]

The casting of cannon and shot at Hope Furnace in Rhode Island, at Salisbury in Connecticut, and in the iron plantations of Pennsylvania and New Jersey marked the entry of the colonial iron industry into the war effort. It was previously noted that, on the eve of the Revolution, the American colonies possessed an iron industry virtually equal in size to that of England. Already in Europe in the eighteenth century, shot, shells, and military hardware, rifle barrels, bayonets, fittings for ships, and carriages were produced by iron technology. Iron also was used to cast cannon, but the favored material for field artillery needed to support armies was brass, not iron, although the more expensive bronze was sometimes used when brass was not available. Iron was cheaper but more subject to corrosion and dangerous failures when fired. To use iron safely, a very heavy casting was felt necessary, making iron cannon difficult to move, hence the preference for the lighter copper alloys that were more suitable for field use. On the other hand, cannon for coastal defense did not have to be moved, and siege artillery depended for its effectiveness more on the great size of shot it could hurl than on mobility. Iron was the material for defensive artillery, and the colonies, with long coastlines and navigable rivers subject to the depredations of the British fleet, were individually on the defensive. The casting of iron cannon, however, presented a series of new problems for the American ironmasters.[27]

There is no evidence to suggest that cannon had been cast anywhere in the colonies prior to the Revolution. The techniques of casting large iron objects such as anchors were known and used, but the two primary problems facing American ironmasters in the casting of cannon were how to standardize the size of artillery and how to make sound castings. The first problem was solved by the adaptation of the English military shot tables. The standardization of ordnance had been begun by the French toward the middle of the sixteenth century, when the calibers of French artillery were reduced to six. By the Revolution, each nation had its own shot tables, which listed the cannon caliber and shot diameter for a given weight of shot. The colonies and the Continental

Congress adopted the English standards which remained in use throughout the war. When, later in the war, the French began to supply artillery, ironmasters also had recourse to the French tables which required a different set of shot weights and diameters.[28]

Revolutionary War Naval Cannon
Colonial iron furnaces cast hundreds of cannon for use by the Continental army and navy. The cannon were similar to these shown here in a reconstruction of H.M.S. *Charon*, sunk during the battle of Yorktown. Almost no cannon from this period have been preserved. (*Courtesy, United States Department of the Interior, National Park Service*)

The second and more serious problem, the production of sound castings for iron cannon, could be solved in practice only by trial and error methods. Once it was felt that a satisfactory cast had been made, the cannon were tested by firing "proof" loads, heavier powder charges than would be used in actual practice. If the cannon did not blow up in the proof, it was deemed sound. The same practice was used by European cannon makers, since the complexities of iron casting were not fully understood until well into the nineteenth century. Not all colonial attempts to cast cannon were initially successful. One of the first two cannon cast by Hibernia Furnace in 1776 passed proof,[29] but Thomas Hughes in Maryland was less fortunate. In the first tests of his castings, four out of five burst, "killing poor Matthews," one of the inspectors.[30] Within a few months, Hughes had mastered the technique of producing sound castings, and he furnished a steady supply of cannon to Maryland and the Continental Congress throughout the war.[31]

The other colonies conducted individual efforts to obtain supplies of cannon and shot similar to those of New England and Pennsylvania. After discovering that "none can be had from any of the Eastern Governments, they are very scarce thro' the whole Continent," Maryland placed substantial orders for cannon with the Hughes brothers at Fredericktown and for shot with John Weston at Baltimore and Charles Ridgely on the Little Gunpowder River in Baltimore County.[32] Nowhere was the resulting shortage of supply more keenly felt than in those colonies where little or no iron production was occurring before the Revolution, particularly in the South. The exigencies of war stimulated the demand for the production of iron in those areas. After its petition to Congress for cannon to defend its towns on the Chesapeake was rejected with the warning, "That the demands of the Continent for cannon are at present of so pressing a nature that none can be spared for the particular use of any state," Virginia authorized the construction of a cannon foundry near Richmond. New furnaces, forges, and shops were built, particularly an impressive complex near Westham, six miles below Richmond on the James River. There, eight iron furnaces, forges, and other facilities were estimated to be able to cast up to three hundred cannon and one hundred tons of shot and shell in a year, although it is unlikely that production ever approached that capacity. Westham performed well for the Continental forces until, on January 5, 1781, British troops destroyed the foundry at the direction of the turncoat general, Benedict Arnold.[33] It was not rebuilt. And North Carolina, without adequate works at the start of the Revolution, attempted to organize one on the Deep River, sending to Pennsylvania for molds and patterns and advertising as far away as New York for an experienced cannon and shot founder.[34] For a brief period two furnaces operated in the northwestern corner of South Carolina, but, as with the Westham works, they drew the attention of the British army during the southern campaign and were destroyed.[35] Throughout the war, therefore, the southernmost colonies, hampered by the lack of existing facilities and skilled craftsmen, were forced to rely on imported materials or "loans" from other colonies, particularly Virginia.

The renewed activity in the production of iron in Virginia and the Carolinas did not survive the war. Only in the valley of Virginia, the gateway to Kentucky and Tennessee territory, was there a lasting increase in iron manufacture during the war. That, however, made no contribution to the war, having been created to supply the growing movement of people westward, crossing the mountains to escape the war.

The great concern for local defense by the individual colonies preceded any actions by the Continental Congress. Throughout much of the Revolution they continued to ignore the efforts to conduct a unified war effort. The presence of stores of cannon and shot in the colonies to some extent relieved Congress of the necessity to provide for "the common defense," leaving it freer to devote attention to offensive warfare. More important, the early conversion of many local furnaces and, as we shall note, the erection of new works to supply war materials at the instigation of the "independent" states greatly facilitated the task of obtaining supplies for the national program, supplies that were vital to the successful prosecution of the War for Independence.

When the Second Continental Congress met in Philadelphia on May 10, 1775, shortly after the battles of Lexington and Concord, the mood of the delegates was increasingly militant. Among them were men who knew from experience the capabilities and importance of the iron industry. At least six delegates who were later to sign the Declaration of Independence had direct involvement in the manufacture of iron at the start of the Revolution. Charles Carroll of Maryland was a shareholder and active participant in the Baltimore Company. Philip Livingston's father owned the Ancram Ironworks at Livingston Manor in New York. Stephen Hopkins of Rhode Island was a partner in the Hope Furnace. George Taylor, ironmaster at Durham Furnace, George Ross, part-owner of Mary Ann Furnace, and James Smith, operator of a forge in York County, all signed for Pennsylvania.[36] George Washington of Virginia, whose father had been one of the founders of the Principio Company in Maryland, also was present initially, representing Virginia.

Yet a climate of conciliation still prevailed, a fading hope that King George III would grant the redress of colonial grievances, averting full-scale war. Congress was slow to attempt to mobilize the resources of the country as long as that hope remained. Early in the summer, however, Congress finally began to make the preparations necessary to wage war. In so doing, it copied the steps already taken by the more pragmatic provincial legislatures and their committees of safety. The first tentative move was the resolution to break its own nonimportation agreement by offering to trade the produce of the colonies with "anyone" who would supply them with powder, muskets, and brass field artillery.[37] Other actions quickly followed. A call was sent out to all the colonies to put the gunsmiths to work making firearms for the Continental forces.[38] Because, at the start of the Revolution, all firearms were in nominal possession of the colonies or individuals, early calls for recruitment usually included a requirement that men supply their own arms. As late as March 1776, Washington was instructed to discharge troops whose states could not provide them with arms.[39] Later, the colonies equipped militia from their own stores. Moreover, several colonies already had begun employing gunsmiths to meet their own needs. The request to supply arms for the Continental forces created conflict between the various procurement programs. Eventually the colonies provided a large store of firearms to the Continental effort, but not always with good grace.

As 1776 began, with Washington's army camped before the gates of Boston, Congress appointed a committee to investigate the state of cannon in the united colonies, the needs for defense, and the possibilities for casting cannon, both brass and iron. On the basis of the committee report, in February, Congress authorized its first purchase of cannon, 250 twelve-pounders, 50 nine-pounders, and 52 four-pounders, to be cast in the colonies as soon as possible.[40] The appointment of the cannon committee was the first step in providing badly needed ordnance. Unlike the colonies, however, Congress was less assured that the furnaces and foundries of America could be mobilized to supply all of the necessary artillery. For one thing, copper and brass, the chief constituents of field artillery, were in short supply in the colonies. In March, Silas Deane was dispatched as ambassador to France. His instructions included an urgent plea to obtain clothing and suitable ammunition and arms for twenty-five thousand men, as well as a hundred field pieces.[41]

By modern standards, even those as early as the American Civil War, the role of artillery in the Revolution was limited. This was particularly true in land battles where armies had to move long distances, often through nearly trackless wilderness, making the transport of cannon difficult. Throughout the Revolution, land battles were fought by armies of less than ten thousand men. The use of artillery as a tactical weapon was still in the developmental stages. Emphasis was on the infantry volley with field guns used to support the infantry position and protect it from attack rather than as offensive weapons. Field artillery usually was supplied in proportion to the number of infantry in the ratio of two cannon per battalion of approximately 470 men. The small size of opposing forces served to limit the number of cannon employed.[42] When the British General Burgoyne began the march southward along Lake Champlain in 1778, his field train included eighty cannon, twelve howitzers, and sixty-four mortars. When the army entered the forests below the lake, however, only thirty-six cannon were carried forward to the battle of Saratoga, twenty-six assigned to the different corps under his command, and the other ten formed his artillery park or replacement supply. The rest had been left to reinforce the defenses at Crown Point and Ticonderoga.[43] The surrender of Burgoyne's army yielded barely six thousand prisoners, including several hundred women and children. At the even more decisive victory at Yorktown, the total prisoners numbered less than eight thousand. The Continental forces, often even smaller and hampered by the shortage of supplies, used cannon on an even more limited scale in the field. At the major battle of Monmouth (1778) their main battery consisted of just twelve field pieces.[44]

The largest use of cannon during the Revolution was in the many fortifications built at strategic points along the coast, rivers, and lakes. There, combinations of artillery were assembled—cannon for use against attacking land forces, howitzers, and mortars against ships. In turn, the short-range howitzers and mortars were the principal artillery used in besieging fortifications.

Washington's task of besieging Boston was eased by the capture of sixty good cannon with the fall of Fort Ticonderoga on May 10, 1775. As the Continental noose tightened around Boston's neck in the fall of

1775, Washington dispatched Colonel Henry Knox with a company to haul the guns, weighing about 120,000 pounds, across the Berkshires to Boston, where, together with the train of four field artillery pieces originally owned by Massachusetts, twelve cannon borrowed from Rhode Island, and a fifteen-inch brass mortar captured on the British ordnance brig *Nancy*, they constituted his siege artillery.[45] In 1781, Washington used upward of fifty iron and brass howitzers to besiege Yorktown; Cornwallis, whose army had been equipped for field maneuvers, surrendered seventy-four brass field pieces used to stand off assaults on his defenses.[46]

Although the overall requirements for cannon by the colonial forces were not large, the need still greatly exceeded the supplies available at the start of the Revolution, particularly in the critical area of field artillery. Cannon captured in land battles, especially at Fort Ticonderoga and Saratoga, constituted one of the chief sources of supply for Washington's forces in the first half of the war. The British evacuation of Boston in March provided another windfall for the Continental forces: thirty cannon and two mortars, all of which had been spiked by ramming steel cones into the touchholes, but which were salvageable. There were another 135 pieces at the Castle, the fort defending the harbor, but most of them had been broken beyond repair. While the siege artillery was sent forward to defend against an invasion of New York, the captured cannon were turned over to Massachusetts for her defense. But these gains were offset by substantial losses such as those in November 1776, when, following the rout of Continental forces on Long Island, only one train of field artillery could be salvaged from the loss of Forts Lee and Washington on the Hudson. Cannon frequently failed during firing, creating a steady drain on both men and materials. The failure rate among early cannon was notoriously high and plagued the armies on both sides throughout the Revolution. General Washington lost five guns during one shelling of Boston. Use of cannon was dangerous for the men also. When forwarding two "replacement" artillery officers to New Jersey, Washington remarked, "It is a melancholy Consideration that in these cases we suffer more from our own Artillery than the Enemy."[47]

Since the primary concern of Congress was the conduct of a land war, the supply of field artillery was given greatest attention. Hence the efforts of Congress possessed a different orientation from those of the colonies, which primarily were concerned with defensive weaponry. Brass was specified for field pieces if at all possible; otherwise iron was acceptable. To relieve the shortage of brass, it was proposed that any disabled guns of that metal, such as a number of those at Boston, were to be remelted, and copper and brass of all kinds were requisitioned throughout the colonies.[48] In New York an air furnace was set up shortly after the British evacuation of Boston, superintended by Colonel Knox. There, within a few weeks, James Byers, an experienced founder, was successfully casting five-and-a-half-inch howitzers from remelted brass and copper. Knox noted that for howitzers or cannon up to eight-inch bore, brass was preferred because of its one-to-three weight advantage over comparable iron ordnance. Knox encouraged the appropriation of copper throughout the colonies for this purpose, particularly noting that "there are a great number of Stills which are only a pest to society which ought to change their form."[49] Considering the urgent need for field artillery, it is to the credit of the colonies that there are no records of the remelting of church bells to alleviate the shortage of copper.

The duties of the cannon committee were subsumed by a new Board of War and Ordnance in June 1776, and a foundry was ordered to be set up in Philadelphia to cast brass cannon. James Byers, who was recommended by Washington as already having cast some "good ones" in New York, was hired to supervise it.[50] Early in 1777, Henry Knox, now general of artillery, was dispatched to Hartford to establish a cannon foundry there. Knox proposed the building of the foundry at Springfield and was given Washington's authorization to proceed. Benjamin Fowler went to York, Pennsylvania, on a similar mission, with orders to build shops for forty blacksmiths, twenty wheelwrights, and all other necessary artisans.[51] Throughout implementation of these efforts, objections to them were raised because they often conflicted with the use of craftsmen and materials by the individual states. Moreover, the states responded with urgent petitions to Washington and Congress to dispatch the sorely needed cannon for their own defenses.[52]

Meanwhile, in France, Deane had found a sympathetic ear in the person of Pierre Augustin Caron de Beaumarchais, who immediately arranged the purchase and shipment of twenty thousand fusils and up to two hundred brass cannon and twenty-four mortars. The first of the French supply ships, the *Mercury*, reached New England in March 1777 with twelve thousand muskets. In May, the *Amphitrite* arrived carrying forty-one hundred stand of arms, nineteen mortars, forty-eight brass field pieces, powder, shot, and shell.[53] Because of the start of French arms shipments, the foundry proposed by Knox was not built. Instead, Springfield was used as a major depot for arms and repair in the northern states. The present federal arsenal was established in 1794. The foundry in Philadelphia operated only briefly until the British capture of the city. Byers was sent to York and subsequently to a new foundry established at Carlisle to continue his work. By early 1778, however, General Horatio Gates noted, "As the Number of Cannon on the Continent is considerable, the Board have not yet concluded to set Mr. Byers at work immediately."[54] Although Byers remained in the employ of Congress until August 1781, he frequently complained of inactivity. From 1778 until the end of the war, substantial purchases of arms by American agents in France, particularly of brass cannon, sufficed to fill the critical need for field artillery.

Cannon alone were of little value without the powder and shot to fire them. The extensive efforts to obtain adequate supplies of powder lie outside the scope of the present study, but it was in the area of the casting of shot and shell that colonial ironworks provided one of their most successful services to the conduct of the war. The techniques of casting solid shot and hollow shells were essentially the same as those used to cast stove plates or kettles and were well known to ironmasters. Wooden patterns were shaped to the desired specifications, around them molding sand was tamped into a wooden frame, the pattern was removed for reuse, and the molds were carried to the furnace to be filled from ladles of molten iron. The key craftsman in the process was the patternmaker, who had to ensure correct dimensions and proper roundness for shot molds. Furnaces without the expertise or capacity to cast cannon could and did respond to the calls from the Continental Congress for the casting of shot and shell.

Throughout the war, major contracts were let and continuously renewed with over a dozen of the larger furnaces in New Jersey, Pennsylvania, and Maryland. Returns to ordnance stores showed that by 1780 thousands of tons were being received annually from the larger furnaces, and smaller works, such as King's Furnace near Taunton and an air furnace erected at New Haven to remelt pig, also were supplying shot and shell for Continental forces. This is one area where production from American ironworks was more than adequate to meet Continental needs throughout the war.[55]

Supplies of cannon for siege batteries and coastal defense remained marginal, however, as British campaigns in New York, New Jersey, and Pennsylvania threatened the existence of the colonial cannon foundries. The furnaces in north central New Jersey and southern New York, the region of active fighting, were particularly hard hit. The Forest of Dean Furnace in New York was abandoned when the British captured the nearby Continental forts, and British troops razed the furnace at Mount Holly in their march across New Jersey.[56] Other furnaces and forges were attacked periodically. The danger was not only from enemy intervention; Thomas Lesher of Pennsylvania complained that he was forced to close the Olney Forge and to suspend construction of a furnace when American foraging parties drove off his livestock and cut all his wood.[57] Ironmasters often complained that the slowness in receiving payment from Congress threatened disruption of their operations. In 1780, John Jacob Faesch of Mount Hope Furnace urgently requested funds, noting that "my People are almost Naked for want of Shirting" and that he did not have enough money to pay his taxes, which were coming due.[58] The several units of Peter Hasenclever's old American Iron Company, Ringwood, Charlottenburg, and Long Pond furnaces, all closed by 1778 because they lacked the men to run them. State legislatures tried to assure the operation of furnaces by granting militia exemptions to workers, but it was not always possible to maintain an adequate work force. Moreover, Washington strongly protested the indiscriminate granting of exemptions to ironworkers unless the works were being operated exclusively *"for the public,"* since many furnaces and forges attempted to conduct business as usual, disregarding the general war effort.[59]

After the desolate winter at Valley Forge in 1777–78, American fortunes began to rise. The supply of cannon and other firearms improved through French imports and domestic manufacture. More of the output of the iron furnaces originally exploited for state defenses was directed to the Continental effort as the danger to some states, particularly the northern ones, began to diminish. Still, in 1780, a request to Connecticut to cast cannon was returned quoting such an exorbitant price that Congress balked and tried to suggest a "loan" instead.[60] The need for heavy cannon remained acute, however. For the decisive campaign at Yorktown, in 1781, Washington had to appeal to the governors of Massachusetts and Rhode Island to lend him all available cannon to conduct the siege of Cornwallis's forces. Unlike the lack of cooperation the colonies had displayed during the siege of Boston, this time they sent the cannon with dispatch.[61]

Although supply of the land operations against the British army at Boston was the most pressing concern for the Continental Congress in the summer of 1775, they were aware of a far more imposing threat in the overwhelming naval superiority of the British fleet. That same superiority that had ensured safe-conduct on the seas for American vessels and rising prosperity for her merchants throughout the colonial period now represented a principal means to strangle the growing rebellion. The long and largely undefended coastline of the colonies was nearly everywhere vulnerable to attack from the sea; hence the early concern by the individual colonies for coastal artillery. Furthermore, the British fleet could disembark troops and, in most regions, ensure a steady supply of military stores to keep them in the field. The military factor was only one potential consequence of British naval dominance; another was the threat it posed to American trade. Although during the eighteenth century the American colonies had achieved a degree of self-sufficiency in some areas, such as the production of iron, they still were dependent on imports for many items such as woolen cloth and for some critical materials, copper, brass, and steel being among the most important. Moreover, colonial manufactures tended to be concentrated in the North, and the extensive coastal traffic essential to the economy of the states was vulnerable to naval attack. Because of the need to protect the coast and foreign and local trade from naval depredations, the Continental Congress early began to consider means to counter the threat of British naval dominance.

At the start of the Revolution, a thriving shipbuilding trade existed within the colonies and a sizable merchant fleet operated out of America's port cities. One avenue immediately open to the Americans was the commissioning of privately owned vessels as privateers to prey on British merchant shipping. The states took the lead in commissioning such vessels, which were outfitted by merchants who had the added incentive of sharing in the value of any prizes captured by their vessels.

Nearly two thousand ships were commissioned by the states in the course of the war, the vast majority being small, fast, and lightly armed.[62] Typical of the privateers was the Pennsylvania ship *Commerce*, commanded by Thomas Truxton, who later became an admiral in the United States Navy. Owned and outfitted by Philadelphia merchants, the ship mounted fourteen cannon with a crew of fifty men.[63] The same long coast that made defenses so difficult for the Continental forces made blockades nearly impossible for the British navy. Privateers were able to slip in and out of the bays and estuaries to harass the British merchant fleet in the Atlantic and Caribbean. After the entry of France into the war, privateers were able to operate out of French ports, where they did even greater damage in British home waters. During the Revolution, American privateers captured upward of six hundred British ships and nearly $20 million in supplies, to the great embarrassment of the British navy.[64]

The efforts of privateers also yielded important gains for the Continental forces at a time when no navy yet existed to challenge the British fleet. In November 1775, a privateer captured the ordnance brig *Nancy* on its way to supply General Gage's besieged forces at Boston. Among the stores were over two thousand stands of arms, a number of iron and brass cannon, and a huge fifteen-inch brass mortar, dubbed the "Congress," which was used by Washington in the siege of Boston.[65] The capture came at a time when the shortage of cannon in the colonies was most acutely felt.

On October 30, 1775, the Continental Congress took the momentous step of authorizing the creation of a navy with the appointment of a committee to devise "means to furnish colonies with Naval armament." It was proposed to build thirteen frigates, of from twenty-four to thirty-two guns, one each by Maryland, Connecticut, and New Hampshire, two by Massachusetts, Rhode Island, and New York, and four by Pennsylvania. The Congress fully anticipated that all materials necessary to outfit them, with the exception of cannon and an estimated one hundred tons of powder, could be supplied in the colonies.[66] Even while plans were going forward to build Continental frigates, a small fleet of converted merchantmen, led by Commodore Esek Hopkins, conducted a surprise raid on British installations at New Providence and Nassau in the Bahamas, carrying off eighty-eight cannon and fifteen mortars.[67]

Although pressed forward, the building of the frigates encountered a series of problems, not the least of which was the chronic shortage of cannon in the early years of the Revolution. Most of the guns taken by Hopkins in the Bahamas were, in the absence of proper carriages, suitable only for coastal defense. A number were installed at Groton to protect the entrance to New London harbor. Two dozen others were deposited with North Carolina, a state devoid of cannon or the facilities to make them in the early years.[68] Pennsylvania, with a thriving iron industry, had little trouble supplying iron for the frigates being built there. By May 1776, Colonel Mark Bird at Reading Furnace was ready to prove cannon he was casting for the fleet from patterns designed by David Rittenhouse, and Durham Furnace had contracted to supply three hundred tons of pig iron for the ballast for two frigates.[69] Similarly, in New York, Orange Furnace had supplied fifty tons of ballast by early March for the frigate being built at Poughkeepsie and had orders for tons of grape shot, cannonballs, and twenty nine-pound cannon for the ship.[70]

In New England, however, the rugged spirit of independent action displayed by the states at the start of the Revolution plagued the construction of ships. Hope Furnace in Rhode Island engaged to cast ordnance for the Providence frigates, including twelve eighteen-pounders, the largest naval guns cast in the colonies, but, when requested to supply guns for the *Raleigh,* being built in New Hampshire, which lacked furnaces, claimed that the commitment could not be met. Nor would the committee responsible for building the Rhode Island frigates release cannon already cast to outfit the *Raleigh* although its construction was far advanced over the Rhode Island ships.[71] Massachusetts agreed to supply cannon for the frigate *Boston* out of its stores but perversely directed its agent that the cannon were to be redelivered to the state unless Congress agreed that the frigate would cruise the coastal waters "to protect the trade of the State."[72] A significant factor contributing to the failure of agents to obtain the cooperation of New England ironmasters in casting cannon for Continental frigates was the much higher profit that could be obtained by casting guns for the use of privateers. In vain, agent Thomas Cushing lamented, "O for Cannon! The spirit of Privateering prevails so amazingly here [Boston] that Cannon cannot be procured, if at all, but at a most extravagant price."[73]

Eventually, nine American frigates were completed, but the overwhelming British naval superiority that could effectively blockade the approaches to harbors and bays made the use of American ports untenable as bases for such large vessels. That superiority more than the problems of supply brought to an early end efforts to construct a navy on American shores. The output of furnaces that had supplied the construction of the frigates was directed to casting ordnance for Continental land forces.

In the absence of an adequate navy, forts and cannon were the normal means used to discourage the use of American harbors and rivers by British ships. Defense of the Hudson River, however, required a unique contribution to the war by the iron industry. In October 1775, a committee of John Morton, Robert Livingston, and Silas Deane was appointed to investigate ways to make the Hudson River defensible, beginning a sequence of events that ended in the casting and forging of the great iron chain at West Point, one of the most unusual metalworking projects of the Revolution.[74] The earliest efforts to close the river to navigation were the use of fire ships against a British flotilla and the fortification of the heights at the northern end of Manhattan Island. Both of these proved ineffectual. In 1777, the Americans attempted to close the river by suspending a chain and boom across the Hudson at Highland, about five miles below West Point. Initially, the function of the iron chain was to hold the floating rafts of the wooden boom in place, the boom constituting the major navigational obstruction. Before the work was completed, however, a clear distinction was drawn between the boom and the chain, and the focus was shifted to the construction of the chain itself as the principal barrier. A chain consisting of 303 clips, 276 links, and 197 bolts and weighing over fourteen tons was installed above Peekskill, protected by batteries at Forts Montgomery and Clinton.[75] The theory was that the heavy chain would stop British ships and prevent them from moving up the river. While stalled against the chain, they would be destroyed by cannon batteries on both shores of the river. A secondary boom of pointed logs was to be floated just below the chain to slow ships and prevent their butting the chain with enough momentum to force a possible break. The use of an iron chain to block waterways in this fashion appears to have been an entirely new concept in the annals of warfare. In October 1777, before the log boom could be installed, British troops landed below the forts, outflanked them, and took them easily. The chain fell to the British, who eventually sent it to Gibraltar to protect the shipping at the breakwater there.[76]

Peter Townsend, the proprietor of Sterling ironworks, immediately was contacted to construct a new barrier chain to be placed in the Hudson at West Point. From February to April, Sterling produced the great links, some three and one-half inches thick and over three feet long. They were hauled twenty miles by wagon to the edge of the river at New Windsor, where they were forged together. On April 30, the great chain, weighing almost 150 tons, was in place. The British never attempted to pass it.[77]

With the opening of hostilities, one metal that had elicited little interest throughout much of the colonial period, lead, suddenly became a material of great importance. England had abundant supplies of lead, and it was mined extensively, primarily for the silver that often was found associated with lead ores. As a result, the lead, which was regarded as little more than a by-product in the extraction of silver, was produced in quantities far exceeding existing demands.[78] It was used as ballast on ships sailing to America and there sold cheaply in the colonial markets. Small amounts were used in the recasting of pewter or in bronze casting and larger quantities for weights on fishing lines and nets or for musket bullets. With the growth of cities and rising prosperity, a new market for lead was found in the making of sash-weights for the windows of town houses.

The early explorers and settlers in America had, on occasion, noted the presence of lead-bearing rocks, and a few mines had been opened in

Great Chain at West Point
Links and pins of the chain used to block the Hudson River to British warships
during the Revolution. This project was one of the most ambitious and
successful contributions of the colonial iron industry to the War for Independence. *(United States Army Photograph)*

the expectation of finding silver. A mine in Southampton, Massachusetts, for example, was operated intermittently from about 1680 to the
eve of the Revolution for the silver content of its ores.[79] Although the
presence of lead ore in the colony of New York was recorded from 1737
onward, the first major attempt to mine it occurred after 1768 in
Westchester County, where Frederick Phillips took out a ninety-nine-
year lease to operate a "silver" mine.[80] Because of its cheap availability
from England, no serious effort was made to find lead or to develop
deposits of that metal in the Atlantic colonies before the Revolution.

Once the shooting started, however, the need for lead to make
bullets became urgent. The colonists no longer could look to England
to satisfy their requirements. As early as August 1775, George
Washington wrote to Governor Jonathan Trumbull of Connecticut and
General Philip Schuyler requesting that a supply of lead captured at
Fort Ticonderoga be forwarded to Cambridge, where the shortage of
bullets was being felt "most sensibly."[81] At the same time, Benjamin
Franklin wrote to the Committee of Correspondence in Albany requesting that some of the same stores be sent to Philadelphia for the use
of the Pennsylvania militia.[82] Franklin's letter was yet another example
of the lack of coordination between the individual colonies and the
Continental Congress in providing for the materials of war during the
early stages of the Revolution.

As in the case of cannon casting, the initiative in obtaining lead fell to the colonial governments. Throughout the colonies, there was a search for new lead deposits, and abandoned lead mines, such as those in Dutchess County, New York, originally another silver prospect, were reopened.[83] Connecticut established a commission to investigate possible mines in the colony and to stockpile lead and lead ore and, if necessary, to build a smelting furnace at state expense. Under the direction of the colony, a vein of ore at Middletown, Connecticut, was opened and worked extensively, producing over five thousand pounds of metal before the vein ran out in 1778.[84] A little later, in Pennsylvania, General Daniel Roberdeaux erected a fort near the lead mines in Sinking Spring Valley and mined lead for the state until scarcity of the ores and trouble from Indians forced him to stop.[85] South Carolina offered bounties for the erection and operation of lead works, and North Carolina took steps to purchase lead from Virginia and explored the possibility of opening a mine in Halifax County.[86] The vast lead deposits along the central Mississippi River, which had been explored and mined by both the French and Spanish in the eighteenth century, were known to the English colonists, but the remoteness of the region and foreign control precluded any consideration of drawing supplies from the West.[87]

The discovery and opening of mines took time, however. To avert an early crisis, the colonial governments adopted several expedients to obtain supplies of lead. The purchase of lead in any form was authorized. Even pewterware, an alloy of lead and tin, was bought and melted to make bullets. In Albany, Philadelphia, and other cities, lead window sashes and clock weights were bought up at prices as high as 6d. the pound, and in Pennsylvania the Supreme Executive Council even ordered the casting of replacement weights of iron to minimize the inconvenience to the inhabitants.[88] Perhaps the most unusual incident in the search for lead occurred in New York City. There, after hearing the first reading of the Declaration of Independence, the citizens retired to the park at Bowling Green where, it was reported, "The equestrian statue of George III, which Tory pride and folly raised in the year 1770, was, by the sons of freedom laid prostrate in the dirt, the just desert of an ungrateful tyrant! The lead wherewith the monument was made is to be run into bullets, to assimilate with the brain of our infatuated adversaries, who, to gain a pepper corn, have lost an empire."[89] The bulk of the gilt statue, one-third larger than life size, was transported to the forge of former Governor Oliver Wolcott in Litchfield, Connecticut, where it was melted down. It was reported that forty-two thousand bullets were produced from the shattered pieces.[90]

All of these efforts, however, were eclipsed by a major lead-mining operation undertaken in the southwestern corner of Virginia that subsequently supplied a large part of the lead used by the colonial forces. The mining of lead on the Great Kanawha River in what is now Montgomery County, Virginia, began in 1759. The discovery of lead ore apparently rich in silver prompted the organization of a company by John Robinson, Governor Francis Fauquier, William Byrd III, and John Chiswell, men prominent in Virginia government and society and, in the case of the Byrd family, prominent in the search for metallic ores. For £2,000 the group bought and began to develop a thousand-acre tract. On Robinson's death in 1768, Edmund Pendleton found himself executor of Robinson's estate but displayed little interest in the mines then operated by a small force of slaves under the direction of Byrd, the only surviving partner.[91] In 1775, however, Pendleton became the head of the revolutionary government in Virginia, and the urgent need for lead prompted him to take immediate steps to expand the working of the mines, which had produced "no profit for ten years past."[92]

The Virginia delegates to the Continental Congress were no less appreciative of their importance. In July of 1776, they informed Pendleton: "We take the liberty of recommending the lead mines to you as an object of vast importance. We think it impossible they can be worked to too great an extent. Considered as perhaps the sole means of supporting the American cause they are inestimable."[93] The Virginia legislature responded to the recommendation by passing a law effectively taking over the operation of the mines in the name of the state.[94] Another law passed in 1777, by which state control was maintained until several years after the conclusion of hostilities, included the mine workers among those excepted from militia duty.[95] In his *Notes on the State of Virginia*, written in 1782, Jefferson described the lead mines as employing about thirty men to produce up to sixty tons of lead per year. A stamping mill and furnace were in operation on the opposite bank of the river a short distance away. The principal drawback to their operation was the location, which required an overland transport of 130 miles to the James River.[96]

When the main theater of war shifted to the southern colonies in 1780, the strategic importance of the lead mines, as the only major source of supply for the South, vastly increased. A Tory uprising in that year had as one of its principal objectives the capture of the mines. Earlier, Tories had incited attacks on the mining settlement by hostile Cherokees. Virginia had found it necessary to dispatch militia contingents to the mines and, finally, to maintain them there to protect the vital supply of lead.[97] Although threatened by the uprising, work at the mine was not interrupted. From the operation of the mines, Virginia continued to supply lead for its own militia and those of the states to the south, in addition to the Continental army.[98]

Then, in the spring of 1781, on the eve of the decisive campaign against General Charles Cornwallis's forces, a new crisis developed when the main vein at the mines seemed about to run out. Jefferson urged the employment of additional hands, including slaves, to expand the diggings. As the output continued to dwindle, he appealed to Congress for assistance, importuning the Virginia delegates: "It is impossible to give you an Idea of the Distress we are in for want of Lead. Should this Army from Portsmouth [four thousand British troops] come forth and become active (and we have no reason to believe they came here to Sleep) our Affairs will assume a very disagreeable Aspect."[99]

By May, it seemed the crisis had passed, for the mine superintendent, David Ross, was able to report the expected production of "40 to 50 ton made in a Short time"; about seven thousand pounds were on hand.[100] Although the output from the mine did not reach Ross's optimistic prediction, just under thirty tons were produced in 1781, more than sufficient to satisfy the increasing demands by the army during the summer.[101] Whereas the lack of five tons of lead might have spelled defeat for the colonial forces, the supply did not falter, and the southern campaign culminated in Cornwallis's surrender at Yorktown on October 10.

At the start of the Revolution, the smelting, casting, and forging of iron was, without question, the most important metal technology employed in the colonies. Of growing significance was the fabrication of iron and other metals requiring the specialized techniques of the metals craftsman. The iron industry, as we have seen, achieved new and greater importance supplying the materials of war. The status of the colonial metals craftsman during the Revolution, however, is difficult to determine.[102] Large numbers certainly exchanged their profession for that of soldier in the Continental forces. A few, such as John Fitch, a coppersmith and later steamboat inventor, took no active part in the Revolution, or found other pursuits to tide them over the disruptive period. When Fitch's shop at Trenton was burned by the British in 1778, he drifted westward to the Ohio, and engaged in land speculations. By the end of the war, he was a prisoner of hostile Indians.[103] Yet others, such as the silversmith Charles Bruff of New York, took advantage of the British occupation of the city to carry on a lucrative trade with the troops.[104] For many craftsmen, however, the Revolution appears to have been a lean time. The supplies of raw materials were interrupted, especially copper, brass, tin, and silver imports, or diverted to the war effort. Transportation was uncertain and markets were adversely affected by the ebb and flow of battle throughout the colonies.

Of all the metals craftsmen, blacksmiths and gunsmiths were the least affected by the war. The skills of the blacksmith still were vital to the maintenance of the ongoing agricultural society and found new outlets in the support of the army. Washington repeatedly called for the training of blacksmiths in the ranks as armorers, and in the campaigns after 1777 found it expedient to equip each brigade with a traveling forge to satisfy the immediate needs for smithwork.[105]

One unusual feature of metalworking during the Revolution was the erection of air furnaces in or near many of the major cities. These cupolalike structures were charged with pig or scrap iron to produce a wide variety of castings. Air furnaces in New Haven, New York, Philadelphia, and Richmond cast stove plates, kettles, salt pans, and a wide variety of other wares for local populations cut off from their normal sources by the disruptive effects of the war. In New York, as already noted, the air furnace was used to cast brass cannon before the British capture of the city. In Philadelphia and Richmond, air furnaces also cast heavy forge hammers, rolls for slitting mills, and blacksmith's anvils, items essential to the prosecution of war in America, but hitherto imported from England. The maintenance of the air furnaces, which had large components of iron plates and hardware in their construction, called on the skill of the urban blacksmith.[106] Wheelwrights also found ready employment in the outfitting of supply wagons and the making of an increasing number of caissons and carriages for field artillery.

As the war progressed, a number of artisans were employed by the armories and manufactories established by the Continental Congress and several of the states, particularly Pennsylvania and Virginia. At the Continental armory at Carlisle, gunsmiths repaired and maintained muskets and swords, coppersmiths cast belt and shoe buckles, made surveying instruments, and shaped butt and trigger plates for firearms. Wheelwrights plated wheels and ground cutters for boring cannon, while blacksmiths made a broad range of tools from axes to spades, nails, horseshoes, andirons, and padlocks.[107]

The demands of the war imposed hardships on metals craftsmen, but also inspired a series of inventions. In 1777, the Board of War contracted with a Samuel Wheeler for iron cannon of "a new construction" which he claimed to have invented. After a successful demonstration, he was awarded £800 to manufacture more.[108] The same year, John Belton was asked to make or alter a hundred muskets to his design by which, he claimed, "a common small arm, may be mov'd to discharge eight balls one after another, in eight five or three seconds of time," to the great surprise of the enemy, an intriguing reference to a repeating firearm in an age of single-shot weapons.[109]

The gunsmith became one of the most important craftsmen as a result of the war. Congress and the states vied for his services. Pennsylvania appointed Benjamin Rittenhouse to superintend a gun manufactory employing many black- and whitesmiths "who could be taught the art quickly." An estimate of the equipment to be required included three barrel forges, three lock forges and tools, a casting shop for brass founding, forges for springs, bayonets, and small parts, a grinding and polishing mill, and the use of forty men as lock filers, each specializing in the filing of a different part of the lock.[110] The musket was the preferred firearm, but at one point the gunsmiths of Lancaster expressed reluctance to fill a musket order for Pennsylvania because the production of rifles could bring a better price.[111] Muskets and rifles were in such demand that the committees of safety maintained detailed correspondence dealing with their procurement and distribution down to consideration of single weapons. Prices for firearms manufactured within the states varied from £3 to £4 each in the North, where gunsmiths were more available, to the premium of £6 offered by North Carolina.[112]

Some areas felt the need for weapons acutely, as the following letter from the Massachusetts Council to New Hampshire attested: "The Inhabitants of our Frontier Towns on Connecticut River are sending their Committees in the most pathetic manner, begging them to be supplied with fire Arms as half of them (they say) are destitute, and other parts of the State not much better stocked. . . . We must again repeat our solicitation in the most urgent manner to our Sister State to sell us some of the large Quantity of Guns they have lately imported, or a considerable part of our Militia must remain unwilling spectators of the War in which they would gladly assist their country."[113] It is not known whether the firearms requested were supplied, but the letter indicates that, as in the case of cannon, the needs of the army and states could not be met by domestic production alone. Once again, a part of the problem was created by the divergent demands of states and nation for the services of those craftsmen but, in addition, the best efforts of the American gunsmiths were not adequate without appreciable imports of firearms from foreign sources. In 1778, Congress recognized this problem and authorized the erection of a musket factory somewhere in the states to turn out a hundred thousand stand. The factory, to be constructed in accordance with the terms of a proposal submitted by Peter Penet of France, was to include three or four furnaces and 143 workers under the guidance of French master craftsmen. Penet's memorial promised that ten thousand muskets fully equipped with bayonet, ramrod, and cartridge tin and sixty to eighty covered wagons per year were to be delivered by the "factory."[114] If true, that production would have exceeded by nearly double the capabilities of all of the gunsmiths then employed in the colonies working together. But nothing ever came of the great plans because, the mercurial Penet subsequently reported after returning to France to recruit craftsmen, "It is forbidden to give passage to any workman employ'd in such manufactures."[115] In the absence of such a marvelous production facility, the gunsmith continued to be considered one of the most valuable of colonial craftsmen during the war.

The interactions of the revolutionary war with the metals industries and crafts in America were many and complex. The war had been fought back and forth over the states containing the greatest concentrations of metals manufactures. The iron furnaces and forges had played a significant role in supplying ordnance and other war materials to the states and to the national forces, but a price was paid for those services. Some works had been destroyed, others idled by a lack of manpower. The efficiency of several furnaces—Salisbury, Reading, Warwick, Catoctin—was impressive, but there were no significant changes in the technology of metal production during the revolutionary period. Finally, some of the metals crafts also suffered from lack of manpower and material and the disruption of their normal commerce. Before the Revolution, metal manufacture had helped to provide a base for the evolution of an American culture. In contributing to the victory of the Continental forces, the metal technologies helped to ensure the opportunity for the political experiment now beginning.

With the coming of peace, the problems of war were exchanged for the problems of building a new nation. England no longer was an automatic source of supply to fill colonial deficiencies. There was greater need to find new metal resources and to develop the old ones. In the next two decades the struggle for independence shifted from the battlefield to the marketplace.

The Critical Years

The manufacturers of . . . [iron] . . . are entitled to preeminent rank. None are more essential in their kinds, nor so extreme in their uses. They constitute in whole or in part the implements or the materials or both of almost every useful occupation.— Alexander Hamilton, Report on Manufactures, *1791*[1]

Throughout the colonial period, the introduction and growth of metals technology in America drew upon and strongly benefited from the association with Great Britain. The persistent attempts to regulate the growth of manufactures, usually coincident with periods of economic depression in the corresponding British industries, actually had done little to check the development of metal manufacture in the colonies. Laws designed to control or direct the nature of the use of metals were passed, the enumeration of copper and iron and the Iron Act of 1750 being perhaps the most conspicuous examples. Throughout most of the colonial period, however, such laws were poorly enforced. The colonists, moreover, resisted British regulatory attempts. Almost equally important to the development of metals technologies in America, however, was the light emphasis placed on limitation of colonial manufactures in the overall British mercantile philosophy.

The mercantile system, as it evolved throughout the eighteenth century, had as its goal the increase of exports while decreasing imports in order to preserve and enlarge the quantity of specie within a country. Within the system, colonies came to be considered as a source for raw materials and as a market for manufactures. Nominally, therefore, manufacturing within the colonies should have been strongly discouraged.[2] Great Britain, however, with her nearly total control of the seas, preferred to depend on the regulation of trade to achieve her economic ends throughout most of the period. Thus, as Adam Smith observed: "Though the policy of Great Britain with regard to the trade of her colonies has been dictated by the same mercantile spirit as that of other nations, it has, however, upon the whole, been less illiberal and oppressive than that of any of them."[3] America, in consequence, enjoyed a privileged position in her commerce with England, including access to capital, manufactures, and an unregulated flow of skilled craftsmen with their knowledge and experience.

Independence, however, upset that favored position and replaced it with one of competition, not only with Great Britain but with the commerce and industry of all of Europe. To a limited extent, that competition even extended to the industry of the separate colonies, now self-proclaimed states. During the decade of the 1780s, the commercial superiority of the European nations, and particularly that of Great Britain, began to assert itself in the American marketplace. An unchecked flow of durable goods drained away a reservoir of specie accumulated during the war, depressed domestic manufactures, disturbed the commerce of the states, and jeopardized the repayment of state and national debts. The United States were in danger of falling into an economic bondage more severe than any the colonies had experienced, and under the weak Articles of Confederation the federal government was nearly powerless to prevent it.

Metal manufactures, weakened by the effects of war, felt the challenge of foreign competition most directly. The 1780s were a period of relative stagnation for the iron industry and the practice of metal crafts. During the first critical years of the new nation, the role of metals in American society had to survive an ideological reappraisal as well as economic challenge. Intermingled with the political experimentation in the United States was yet another utopian concept, that of a society grounded on agrarian principles, with free trade to all nations. Thomas Jefferson was the most articulate spokesman for the agrarian ideology, which would have relegated the production of metals, indeed all manufactures, to a secondary and limited status.

The same decade that saw the rejection of the confederation concept of government, which gave way to a single central government embodied in the terms of the new Constitution, saw a pragmatic rejection of Jefferson's agrarian dream. The calls for free trade and the preferential status of agriculture in America were superseded by demands originating within the individual states for the protection and encouragement of domestic manufactures. One of the first acts of the new government created by the adoption of the Constitution was the drafting of tariff regulation to achieve the latter ends. The tariff debates in 1789 and Alexander Hamilton's *Report on Manufactures* delivered two years later both emphasized the importance of the domestic manufacture of metals. The decision to encourage that manufacture made in Congress and strongly endorsed in Hamilton's *Report* constituted a milestone in the history of American technology. Although that decision did not immediately command widespread acceptance among all ranks of the populace, and some industries and metals crafts still were dependent on the importation of raw materials, the colonial era of metals in America was coming to an end. The last decade of the century saw the first faltering moves to expand the manufacture of metals in accordance with the philosophy espoused by Hamilton and his assistant, Tench

Coxe. Metals henceforth would have an assured and increasingly important role in the future of the United States.

Following the War for Independence, the United States had hoped for a favorable trade treaty with Great Britain. Although conceding political independence to the United States, the British refused all overtures to recognize economic independence until the negotiations of the Jay Treaty in 1794. They conceived the American position to be one of weakness, capable of exploitation for years to come. John Lord Sheffield, a spokesman for British commercial interests, wrote a lengthy analysis of the potential effects on their trade to be expected from the "independence" of the former colonies. He contended that the future would find the United States bound to Great Britain commercially more closely than ever since "it will be a long time before the Americans can manufacture for themselves . . . [and] no American articles are so necessary to us as our manufactures, &c. are to the Americans."[4] The weakness of the American trade position, item by item, was considered, and Lord Sheffield repeatedly cited high labor costs in the United States, favoring agriculture over manufacture, as the factor most conducive to maintaining British commercial dominance, particularly in the production of metals. Among the articles listed "in which there will be scarce any competition" were jewelry, plate, copper in sheets and wrought into utensils, lead in pig and sheet, tin in plate, bar steel, and "iron and steel manufacture of every kind." In consequence, he concluded, "the solid power of supplying the wants of America, of receiving her produce, and of waiting her convenience, belongs almost exclusively to our own merchants."[5]

Lord Sheffield's contentions concerning the weakened economic position of the American iron industry in the postwar years received striking confirmation when in 1785 Samuel Gustav Hermelin, a Swedish mining engineer, toured the major iron-producing regions of the United States. The subsequent report of his visit contained a detailed description of the ore beds, furnaces, even the variations in the wages, transportation fees, and money rates affecting the price of iron in the American market. Hermelin's observations throw considerable light on the state of American iron production in the 1780s. Foreign travelers, particularly before the war, had often remarked on the number and vitality of the furnaces and forges in America, but none had greater qualifications to make a definitive study of the industry. Hermelin had devoted twenty years to the mining industry in Sweden and in 1782 embarked on a tour of Europe to examine the principal mining regions of France, Germany, Hungary, and England and to visit the allied commercial centers and colleges of mines. While in Paris, he received instructions from King Gustavus III to go to America to investigate the iron industry with a view to improving the commercial relations that had grown up between the United States and Sweden during the Revolution.[6]

Hermelin found that the effects of the war on the American iron industry were extensive and had increased its vulnerability to foreign competition. Furnaces lay idle, a number of facilities had been destroyed, and skilled manpower had been lost or dispersed. Although some new furnaces and forges had been built and others had increased the volume of their output, partly offsetting the damage suffered by the industry, the technology had stagnated.

Included in Hermelin's report was mention of factors bearing on a possible decline in the iron industry, a decline only in part attributable to the conditions of war. He described a number of furnaces and forges destroyed or abandoned during the war, in the latter case chiefly because of a scarcity of labor. But other problems adversely affecting the growth of the industry had come into being following the war—lack of timber for charcoal near growing metropolitan areas, exhaustion of the limited eastern ore beds, the high cost of transport, and, particularly in the North, the cost of labor. Hermelin's report noted specific facilities closed as a consequence of those problems: Reading Furnace in Pennsylvania, no timber; Union Furnace in New Jersey, high cost of ore; Mayberry Furnace in Pennsylvania, cost of transporting ore from beds ten miles distant; forges and small furnaces on the Musconetong River in New Jersey, expense of labor.[7] Hermelin estimated that the average cost at the furnace of producing a ton of pig iron in Pennsylvania and Maryland was £4.5 per ton, while in the only other major iron state, New Jersey, it was closer to £6 per ton. At the time, the market price in Philadelphia was between £8 and £9 per ton, although "there is much reason (to suppose) that the selling price of pig iron will hereafter be 7 pounds per ton," he noted.[8] The cost of converting pig to bar iron ranged from about £21 per ton in Pennsylvania and Maryland to over £25 per ton in New Jersey, exclusive of transportation costs. As a consequence of the multiplicity of factors bearing on the production of bar iron, chief among which were high labor and overland transport costs, it was deemed possible to sell Swedish bar iron for a price just below domestic bar in Philadelphia, a condition that persisted throughout the decade.[9]

That both Sweden and Great Britain now felt able to compete in the sale of iron and steel manufactures in the American market may seem a little surprising when one recalls the relative parity of the American and British industries at the start of the Revolution, but the war had seen marked change in the development of iron manufacture in the two countries.[10] The British industry, not subject to the direct depredations of war, had flourished under its stimulus, aided by the increasing adoption of coke and coal to the smelting and working of iron. Both its total capacity and output per furnace had increased. Bining, from the perspective of the twentieth century, estimated that the total annual production of iron in the colonies had reached thirty thousand tons by 1775;[11] Hermelin, a qualified observer on the scene, concluded following the war that the annual output was only on the order of twenty thousand tons, bar and pig together.[12] By any system of estimates, it is clear that the effects of the war had resulted in a substantial loss of productive capacity by American industry, providing ample reason for Lord Sheffield's optimistic conclusion.

Financial problems tended to slow the recovery of iron manufacture, and the decade of the 1780s produced further troublesome developments for the American industry. In the middle of the decade, a recession, beginning in England, drove down the price of iron in America also. From a brief high of $112 per ton in Philadelphia in October 1784, the price of bar iron dropped to $68 per ton by June 1786, remaining fairly stable at that level for two years. That was very near the level for bar iron in prewar Philadelphia, at a time when conditions of higher output and lower competition and demand prevailed. By the end of the decade, the price had moved back only to $80 per ton.[13] Only a fairly constant demand generated in local markets enabled the iron industry to sustain operations at a reduced level in the face of depressed conditions and strong foreign competition. Many of the furnaces abandoned during the Revolution were not reopened, and few new ones were built. Where furnaces were built, they tended to have little influence on the major mid-Atlantic furnace and market situation. Construction was confined to frontier or backcountry regions, places where growing populations created demands that could be satisfied by existing facilities only at great difficulty and expense. Such furnaces included the first built in what later became Vermont, at Fair Haven and Bennington (ca. 1786),[14] and the Era Furnaces in York County, South Carolina, in the upper Piedmont region (1788), the previous works in the state having been destroyed in the Revolution.[15]

The technology of American iron manufacture, as Hermelin had observed, changed very little during the Revolution and postwar periods. Furnace output continued to fall behind that of Great Britain. There, the introduction of pitcoal to smelt and refine iron led to a rapid increase in production. By 1788, over 48,000 tons of coke pig iron were being made while charcoal pig-iron output, although declining, still totaled 13,000 tons. The average output per furnace was about 900 tons per year versus an average of 350 tons in America. Between 1788 and 1790, the Watt steam engine began to be used to power the blast of British furnaces, further increasing their productive capacity.[16] Such technological changes were not adopted in America for reasons that will be considered in more detail later in this chapter.

Certainly, before the Revolution the details of the changing technology in the British iron industry, including efforts to smelt coal with coke, were available to Americans through a free, albeit at times reluctant, exchange of information and the importation of skilled workmen and equipment. To impede the postwar development of the American iron industry, British economic interests dictated that the further transfer of technology be prevented. Great Britain acted to reinforce the commercial advantages enjoyed by her iron manufactures by passing, in 1785, a law prohibiting "the exportation to foreign parts of tools and utensils made use of in the iron and steel manufacture of the kingdom; and . . . the seducing of artificers and workmen employed in those manufactures, to go into parts beyond the seas."[17] Included in the detailed list of forbidden items were all forms of rolls, anvils, hammers, molds, presses, or models or plans of such equipment. The penalty for each offense was a £200 fine and twelve months imprisonment. The bill was one of a series of legislative acts designed to discourage the development of manufactures, both in the United States and on the Continent. The prohibition against exportation of iron and steel tools and equipment was made "perpetual" by the statute 35 George III, c. 38 (1795). The provision forbidding the emigration of skilled artisans continued in effect until 1825. The effect of these laws, which were vigorously enforced, was to foster what Joseph Walker has termed a state of "colonialism" in the American iron industry.[18]

In 1791, Tench Coxe of Philadelphia drafted what was intended as a spirited rebuttal to Lord Sheffield's *Observations*. Discussing the domestic manufacture in iron, Coxe boasted that the United States produced about one-half of the steel it consumed and over one-half of the annual supply of more than four million pounds of nails. Almost the full output of furnaces and forges was directed to manufacture for the domestic markets, so that exports of pig and bar iron were approximately only half those in the years preceding the Revolution, averaging 300 tons per month in the years 1789 and 1790. Coxe conceded, however, that large imports of bar iron were received from the Baltic as well as Great Britain, 1,288 tons "from St. Petersburg alone, in the year 1790."[19] The true meaning of Coxe's "defense," of course, was that the United States was, in fact, dependent upon foreign imports to supply an appreciable portion of her metal needs, in the case of iron to a greater extent than had been true before the Revolution. After nearly a decade of independence, the state of the iron industry in the United States, and to an even greater extent that of all other metals manufacture, tended to bear out Lord Sheffield's predictions. Britannia still ruled in the American marketplace.

The iron industry and the metal trades had been introduced to the colonies because the interrelated problems of growing domestic needs and the difficulty of obtaining adequate supplies from Great Britain had made their presence in the colonies imperative. That presence had helped to foster the growth of a society that finally deemed itself sufficiently independent to reassess the traditional bonds holding it to the empire. The result of that political reassessment, political independence, was threatening to destroy a major part of the material basis on which the survival and continued prosperity of American society depended. Whether American manufactures could survive the depressed conditions and foreign competition of the peacetime period would be determined by the importance the people of the new nation placed upon their preservation. During the decade of the 1780s, the Americans were realizing their political destiny in the formation of the Constitution. At the same time, they made their decision on manufactures.[20]

Technological differences were not the only factors influencing competition in the metals industries at the end of the war. Of equal significance was the American attitude toward free trade. During the colonial period, Americans had chafed under British attempts to regulate their trade and commerce. In the years before the Revolution, feelings grew acutely sensitive as Parliament passed act after act further restricting colonial intercourse with other nations. The final blow was the American Prohibiting Act of 1774 which expressly forbade trade to and from the colonies to any nation except Great Britain. In 1776, the Second Continental Congress added a clause to the Declaration of Independence denouncing the king and Parliament "for cutting off our trade with all parts of the world." The same year, the Englishman Adam Smith echoed the essential American position when he wrote: "To prohibit a great people . . . from making all that they can of every part of their own produce, or from employing their stock and industry in the way they judge most advantageous to themselves, is a manifest violation of the most sacred rights of mankind."[21]

American feeling ran strongly in favor of free trade, unencumbered by tariffs or restrictions of any kind. So deep-seated was this antipathetic feeling toward regulation that, with the exception of Virginia which continued to tax tobacco exports and liquor imports, almost no legislation on trade by the individual states or the nation existed from 1775 to 1781. Trade with Great Britain was prohibited by every state during the Revolution under severe penalties, but even that restriction was lifted after peace was restored.

Such was the American attitude in 1783. When Patrick Henry addressed the Virginia legislature with the words, "Fetter not commerce, sir,—let her be free as the air—she will range the whole creation, and return on the four winds of heaven to bless the land with plenty," he voiced the sentiment of the majority of Americans.[22] The ports of the United States were opened, duty free, to all nations. The nations of Europe, however, including Great Britain, did not reciprocate, instead imposing duties and restrictions on the import of American products.

The results were disastrous for manufactures in the United States. In France, a disillusioned Silas Deane had foreseen the consequences of a free-trade policy on American commerce even before war's end, but an unsympathetic Congress ignored his repeated warnings. If Americans did win independence they would find the markets of Europe closed to them, Deane predicted, and American manufactures could not compete with the superiority of the English.[23] Some even felt that the damage to manufactures was not important. At the same time Deane was crying doom in Paris, John Adams, in Amsterdam, was writing to assure his correspondents that no power in Europe had anything to fear from America: "The principal interest of America for many centuries to come will be landed, and her chief occupation agricultural. Manufactures and commerce will be but secondary objects, and always subservient to the other. America will be the country to produce raw materials for manufactures; but Europe will be the country of manufactures, and the commerce of America can never increase but in a certain proportion to the growth of its agriculture, until its whole territory of land is filled up with inhabitants, which will not be in some hundreds of years."[24]

Adams was presenting another facet of American thought in the 1780s, perhaps the strongest challenge to the survival of all manufactures in America, a strong belief that the future economic strength of the United States rested on the products of its farms and forests. The concept of an agrarian America and of free trade went hand in hand. In the early years following the Revolution, many prominent men including Adams and James Madison were outspoken in their defense of free trade and in their disparagement of manufactures. They shared a common belief that the self-sufficiency of the United States in manufactures was unnecessary. As Thomas Jefferson, the man most commonly identified with the agrarian dream, conceived it, such a goal was undesirable: "While we have land to labor then, let us never wish to see our citizens occupied at a workbench, or twirling a distaff. Carpenters, masons, smiths are wanting in husbandry; but, for the general operations of manufacture, let our workshops remain in Europe."[25]

The free trade and agrarian philosophy, if carried through to a logical conclusion, would have seriously damaged manufactures in the United States, particularly the metals crafts which, except for iron, had the added disadvantage of being largely dependent on imports for raw materials. As formidable as was the array of those supporting free trade, theirs was not the only philosophy regarding the role of manufacture in the United States. As early as 1774, Alexander Hamilton had written: "We can live without trade of any kind. . . . It would be extremely hurtful to the commerce of Great Britain to drive us to the necessity of laying a regular foundation for manufacture of our own, which, if once established, could not easily, if at all, be undermined or abolished."[26] This was a statement of a position he was never to abandon. The opinions of Jefferson and Hamilton regarding the role of manufactures in the development of the United States represent the opposite sides of the argument that was debated throughout the 1780s.

Although the abiding belief in the preeminence of agriculture to American prosperity pervaded most of society in the 1780s, problems during the first few years of independence were sufficient to create second thoughts in many minds about the nature of free trade. The issue was money. Under the Articles of Confederation, the Congress had no power to raise revenues to pay its debts or even its operating expenses. Any revenue bill had to have the unanimous consent of the thirteen states. At the national level, this was a continual crisis because that consent could never be garnered. The states also had sizable outstanding debts from the war. The problems of collecting taxes, securing revenue, and repaying war debts were exacerbated by the outflow of specie used to purchase foreign manufactures. A number of ingenious plans were devised to retire state debts or at least to sustain interest payments, but it soon was evident that a continuous source of income was required. Land taxes were distinctly unpopular. The alternative was for the states to use the power being denied to Congress, the institution of tariffs.[27] One by one the states introduced and passed legislation to tax imports in order to raise the badly needed revenue. Some states, notably Virginia, never went beyond the revenue stage, although even for those the list of articles and amounts of impost grew throughout the 1780s.[28] Men such as Patrick Henry found that the need for income overrode their lofty sentiments for unencumbered trade. By 1787 a chastened Patrick Henry was encouraging, albeit not always successfully, the prohibition of and imposts on a range of British goods.[29]

Blacksmith with Forge and Tools
All colonial manufactures suffered in the decade after the Revolution. Petitions from blacksmiths and other metal artisans encouraged individual states and then the Congress to implement tariff legislation protecting the products of American craftsmen and advancing the spread of domestic manufactures. (*Colonial Williamsburg Photograph*)

In the North, however, the damage done to domestic manufacture from foreign competition elicited a more drastic change. Tariff legislation took on the expressed function of protection and encouragement of indigenous industry. The Massachusetts Tariff Act of July 2, 1785, clearly stated that its purpose was "to encourage agriculture, the improvement of raw materials, and manufactures."[30] The preamble to the New Hampshire Act of March 4, 1786, read: "That the laying duties on articles, the produce or manufacture of foreign countries, will not only produce a considerable revenue to the state, but will tend to encourage the manufacture of many of those articles within the same." A 15 percent duty was levied on a wide variety of articles including all wrought iron and brass. Among the few items specifically exempted from the payment of duties by the ever practical New Englanders were artisans' tools and copper bedwarming pans.[31]

The most complete elaboration of the evolving protectionist philosophy was realized in the passage of Chapter 1188 of the Statutes of Pennsylvania, on September 20, 1785. Entitled "An Act to Encourage and Protect the Manufactures of the State by Laying Additional Duties on the Importation of Certain Manufactures Which Interfere With Them," the preamble declared that the welfare of the state and its people demanded that home manufactures be protected from undermining and destruction, "although the fabrics and manufactures of Europe and other foreign parts imported into the country in times of peace may be afforded at cheaper rates than they can be made here." Duties of 12s. per dozen were laid on reaping hooks and sickles and 15s. per dozen on scythes among manufactured articles and 10 percent on all British steel, slit iron, nail rod, sheet iron, cast iron, wrought copper, brass, and bronze and all vessels and utensils of pewter, tin, and lead. Tin, lead, pewter, brass, and copper pig and plate were admitted duty free, however, showing the dependence of the states on the imports of raw materials for those manufactures.[32]

The wording and provisions of the Pennsylvania Act constituted clear repudiation of the free-trade doctrine, but even that bill did not go far enough for some of the interests in the Commonwealth. In November, the major ironmasters petitioned the state House of Representatives for the inclusion of similar duties on bar iron, excluded from protection in the original act. They claimed that the high cost of labor permitted large imports of foreign iron which had acted "to the Prejudice and Impovrishment of all Persons concerned in the Manufacture of Bar Iron" and that further imports threatened the very existence of this important industry.[33]

Even Virginia, which, as already noted, taxed widely and often, was not immune to the protectionist fever. In January 1787, James Madison, who believed with Adams and Jefferson that the future of America rested with her agriculture, wrote to Edmund Pendleton describing the debate on a bill in the Virginia legislature:

The Senate have saved our commerce from a dreadful blow which it would have sustained from a bill passed in the House of Delegates, imposing enormous duties, without waiting for the concurrence of the other States, or even of Maryland. There is a rage at present for high duties, partly for the purpose of revenue, partly of forcing manufacture, which it is difficult to resist. It seems to be forgotten, in the first case, that in the arithmetic of the customs, as Dean Swift observes, 2 and 2 do not make four; and in the second, that manufactures will come of themselves when we are ripe for them.[34]

Many states, however, did not share Madison's sanguine belief that manufactures should be allowed to develop in their own time. On the eve of the Constitutional Convention, the mood of public opinion, particularly in the northern states, was running strongly on the side of protection and the encouragement of manufactures. Connecticut, in 1788, went beyond tariff legislation, exempting all ironworks from taxation and the workers therein from payment of the poll tax.[35]

The new Constitution, adopted in 1787 to replace the obviously inadequate Articles of Confederation, reserved to the federal government the right to regulate trade, so that when the first Congress met in March 1789 a major topic to be considered was a national tariff law.[36] In early April, James Madison of Virginia, proponent of free trade and agrarianism, introduced into the House a resolution to create a temporary impost system on agricultural imports for the principal object of raising revenue.[37] The next day, Thomas Fitzsimmons of Pennsylvania, holding opposite views, proposed the alternative adoption, almost without exception, of the terms of the Pennsylvania Act of 1785, noting that some of the tariffs were calculated to encourage the products of the country "and protect our infant manufactures."[38] The issue was joined—free trade or protection, the agrarian dream or domestic manufactures. In the subsequent debates, lasting more than a month, the Fitzsimmons plan was attacked on various grounds, but, surprisingly, the validity and constitutionality of the protection principle were not challenged. Throughout April, petitions for relief from foreign competition were submitted by artisans and manufacturers in the cities of Baltimore, New York, and Boston, lending strength to the protection argument. Included in the Baltimore petition, the most specific of the three, was a list of enumerated articles affected by importation. Protection was asked for the manufacture of copper and tinware, brass castings, anchors, bolts, spikes, nails, iron and steel edge tools of all types, wrought ironwork, guns and pistols, bar iron, nail rod, and iron castings.[39] Those opposed to fostering manufactures in America submitted no petitions.

The prevailing attitudes toward protectionism became most manifest as the debates focused on an item-by-item consideration of the articles to be taxed. When a tariff on the import of steel was proposed, Richard Bland Lee of Virginia and Thomas Tucker of South Carolina spoke in objection, claiming that the consumption of steel was very great and essential to agriculture, but no facilities existed to manufacture it in the South. Tucker added that such a tax was little more than a tax on agriculture. The issue was not the preeminence of agriculture over manufactures, but the availability of steel. George Clymer of Pennsylvania countered that although the manufacture of steel in the United States was still in its infancy, it already was achieving some success (a Philadelphia furnace had made three hundred tons in two years). Therefore it was "prudent to emancipate our country from the manacles in which she is held by foreign manufactures."[40]

The prevailing mood in Congress supported Clymer. In the final form of the bill that passed both houses and became the first national tariff law, a duty of 56¢ per hundredweight was set on imported steel. Although the 10 percent general duty in the Pennsylvania Act was reduced to 7.5 percent in the Tariff of 1789, the general provisions and the list of enumerated articles followed it closely. A clear consensus had emerged in the nation favoring the encouragement and protection of domestic manufacture, prominent among which were the metals industries.[41]

Through the 1780s the one national figure whose name was most prominently associated with the efforts to encourage the growth of manufactures in America was Alexander Hamilton. Shortly after the passage of the Tariff of 1789, Congress instructed Hamilton, as secretary of the treasury, to prepare a plan for the encouragement and promotion of manufactures sufficient to render the United States independent from foreign sources, particularly for military supplies. Much of the information on domestic facilities was compiled by the undersecretary, Tench Coxe of Philadelphia, who, even more than Hamilton, favored the fostering of domestic manufactures. Coxe also wrote an initial draft which Hamilton edited. The result was the *Report on Manufactures*.[42] In it there was proposed a sweeping system of inducements, including protective tariffs, bounties for new industries, premiums on inventions, and the exemption from duty of raw materials.

After a careful rebuttal of the main argument against furthering manufactures in the United States, including a detailed consideration of the relative role of agriculture espoused by Jefferson and other agrarians, the advantages deriving from the various devices to encourage them were set forth in exhaustive detail.[43] The key to Hamilton's and Coxe's position was the sentiment that "without contending for the superior productiveness of Manufacturing Industry, it may conduce to a better judgment of the policy . . . [to] evince that the establishment and diffusion of manufactures have the effect of rendering the total mass of useful and productive labor in a community, *greater than it would otherwise be.*"[44]

The concluding section of the report dealt with the specific manufactures to be encouraged, beginning with iron, which, it was stated, was "entitled to preeminent rank." Since the facilities for production already were so extensive and its worth universally acknowledged, the report proposed only the increase of existing duties, particularly on steel and nails. Copper and lead manufacturing presented different problems. No copper mines were being worked profitably, although smiths were plentiful. To assist the trades, it was proposed to exempt all forms of raw materials while taxing imported manufactured goods at a high rate. The lead mines in southwestern Virginia still operated, but manufactories were scarce. Additional duties on pewterware were recommended as a further encouragement.[45]

The need for manufactures was, to the authors of the *Report*, more than a means of achieving economic independence from European influences. Moreover, the establishment of manufactures was not to be made at the expense of the undeniably basic agricultural interests in America. Those interests clearly would remain in the forefront of American economic development for the foreseeable future. Rather, manufactures, especially those involving metals, were a necessary adjunct to agriculture as well as a benefit to society as a whole. The growth of America without manufactures would be far slower, its continued political existence more tenuous. In the *Report on Manufactures* there is the distinct echo of the words of Sir Thomas More, "Men can no more do without iron than without fire and water."[46] America must rely on its agriculture, but, as Hamilton and Coxe emphasized, it must have its iron and other manufactures also. Many Americans already were coming to share that perception of America's future.

The threat to the existence of American manufactures by the turbulent economic events of the 1780s had fostered an awareness of their importance and of the need to ensure their continued development. The protection and encouragement of manufactories could not, alone, transform the United States into an industrialized nation, but by adopting the principle of protectionism, an important precedent for the future had been established.

When Alexander Hamilton set out to write the *Report on Manufactures*, the metals industries had already begun a slow but steady recovery from the depressed conditions of the mid-1780s. This was most apparent in the manufacture of iron. Hamilton and Coxe corresponded widely in an effort to solicit information on the extent and amount of existing manufactures as background to the report. Although far from complete, the response to those inquiries convinced Hamilton that iron production was "far more than is commonly supposed."[47] The industry was expanding in its old locations and extending into new areas.

The pattern of growth after 1790 reflected to some extent the protectionist policies employed by the states during the preceding five-year period. Few new furnaces had been built in staunchly protectionist New England, except in Vermont, but forges continued to spring up, especially along the mountainous western borders of Massachusetts and Connecticut. Joseph Cook of Danbury, Connecticut, wrote to John Chester, who was compiling information for Hamilton: "Four years ago there was not a single Forge for [manufacturing bar iron] in this County, and I believe never was; within that time eight or nine Iron Works have been erected, in most of which there are two Forges producing upon a mean about twenty tons of bar iron annually."[48]

In New Jersey, which had not passed protective tariff legislation, different circumstances prevailed. The iron industry of northern New Jersey remained depressed, although a nail manufactory was built at Ramapo near the end of the decade.[49] In 1789, Coxe had reported that there were eight furnaces and seventy-nine forges operating in the state, but that those represented a decline of one-third in the number of furnaces active before the Revolution.[50] During the 1790s, when the provisions of the national Tariff of 1789 afforded a degree of protection to the industry in the state, at least four new furnaces were built in south and central New Jersey, including Martha Furnace (1793) by Isaac Potts of Philadelphia, who constructed a four-fire forge and slitting mill nearby.[51] More pronounced evidence of recovery could be seen in Pennsylvania, the heartland of the early iron industry, where Bining estimated that over seventy furnaces, forges, and slitting mills were built between 1790 and 1800 versus only sixteen in the Confederation period.[52]

The increase in the number of furnaces and forges signified a renewed vitality, yet no change in the technology of iron manufacture accompanied the wave of expansion. Charcoal smelting and water-powered blast still were used, giving the new facilities approximately the same production capacities as the older furnaces. Significant technological developments in the production of iron during this period were largely ignored in America.

One improvement involved a change in the method of converting iron to steel. It was an English innovation, originated in Sheffield about 1740 by Benjamin Huntsman. The process consisted of heating together iron and high-carbon bars in a sealed crucible to a temperature sufficient to melt the metal. The final composition had a uniform carbon content below 1 percent, typical of fine quality steel. Huntsman's method greatly facilitated the production of steel in quantity. But the crucible process was not copied by the Americans, who continued to make steel in limited quantities by the older cementation process, referred to by colonial ironmen as "the German method."[53]

The most important change in iron smelting was the introduction, also in England, of coke to replace the traditional fuel, charcoal. Abraham Darby initiated the technique of coke smelting at Coalbrookdale in 1709, and throughout the century the practice slowly gained acceptance in the mother country, where wood for charcoal was becoming ever more scarce. Its widespread adoption after 1760 permitted a great expansion in the size and capacity of furnaces. English iron production rose rapidly from approximately thirty thousand tons per year in 1775 to over eighty thousand tons per year by 1790. In America, however, no effort was made to attempt coke smelting during the eighteenth century.[54]

Table 1.
Comparative Production of Iron in America:
1770 and 1970

	1770	1970
Population	2,148,076[a]	204,800,000[b]
Bar iron production	21,000 tons[c]	
Pig iron and raw carbon steel production		223,727,000 tons[d]
Annual per capita consumption	.01 tons	1.09 tons
Total employees, basic iron (approx.)	8,000[e]	872,000[f]
Percent population employed in basic iron production	.34	.43

[a]Source: "Estimated Population of the American Colonies: 1610–1780," U.S. Department of Commerce, Bureau of the Census, *Historical Statistics of the United States, Colonial Times to 1957* (Washington, D.C.: U.S. Government Printing Office, 1960), p. 756.

[b]Source: "Population and Area: 1790–1970," U.S. Department of Commerce, Bureau of the Census, *Statistical Abstract of the United States, 1971* (Washington, D.C.: U.S. Government Printing Office, 1971), p. 5.

[c]Estimate, based on Bining's estimate of 82 operating furnaces and 175 forges operating in the colonies in 1775; for 1770, 140 forges (80 percent of 1775) producing 150 tons per year. Blast furnace production is not included, assuming major part of output converted to bar before use (Arthur Cecil Bining, *British Regulation of the Colonial Iron Industry* [Philadelphia: University of Pennsylvania Press, 1933], p. 122).

[d]Source: F. E. Brantley, "Iron and Steel," U.S. Department of the Interior, Bureau of Mines, *Minerals Yearbook, 1970,* 3 vols. (Washington, D.C.: U.S. Government Printing Office, 1972), 1: *Metals, Minerals and Fuels,* 593.

[e]Estimate, based on estimated work force, furnace and iron plantations, all labor, mining, charcoaling, handling, casting—66 operating furnaces at 100 workers per furnace; 140 forges at 5 workers per forge.

[f]Includes 629,000 employed in blast furnace and basic steel production, 226,000 in iron and steel foundries, 17,000 in iron ore mining. Sources: "Nonagricultural Employment, by Industry," *Statistical Abstract, 1971,* p. 219; F. L. Klinger, "Iron Ore," *Minerals Yearbook,* I:583.

If, as I have maintained in the present study, the transfer of metals technology was a necessary concomitant to the growth of civilization in America, it is logical to consider why major changes in the technology, particularly those involving dramatic increases in production, were not copied. A consideration of the state of the iron industry at the beginning of the revolutionary period, when the production changes were being adopted in England, indicates several reasons why the changes were not incorporated in the American technology. First, with the exception of the deep South and along the frontier, both deficient in furnaces, there was no recognized need to expand the capacity of the industry. The data in Table 1 illustrate this situation. By 1770, it is highly probable that American furnaces and forges were producing over twenty thousand tons per year of both pig and bar iron. Since much of the pig iron was converted to bar iron in the forges, the single figure of twenty thousand tons of bar iron (which would be somewhat lower than pig iron production) may be taken to indicate consumption that, for the total colonial population, would have constituted an average of twenty pounds per person per year. That estimate seems the more reasonable when one considers that the annual exportation of several tons of pig and bar iron to Great Britain was offset by the importation of a like quantity of British manufactured ironware, particularly in the colonies below the Chesapeake. An identical consumption figure for Great Britain, with a population of approximately ten million people, would have required a production of one hundred thousand tons of iron per year, over three times the prerevolutionary capacity of the British industry. Technological change to increase production in Great Britain was sparked by an inadequate supply, but no corresponding shortage existed in America.

Moreover, already in 1770, the rapid expansion of the industry in the colonies was beginning to exceed the ability of the American market to absorb its output, and a mild recession resulted. By 1773, expansion again was under way. Very little of it was in the easternmost areas, where surplus iron could have been exported. Rather, new furnaces were built farther west in Pennsylvania and in the Virginia Piedmont to serve a population moving southward and westward.

After the Revolution, several factors combined to block technological change in the iron industry. Both the ravages of war and the economic depression that followed acted to curtail investment in the repair of older furnaces or in the building of new ones. The flood of British manufactured goods, including some iron, glutted the market for several years, making the building of new ironworks with greatly expanded capacity even less attractive. The British decision to bar the export of tools and the emigration of skilled workmen probably was only an incidental factor, since the new British production processes had been fully developed before the Revolution when ample knowledge of them would have been available to the colonists. The wide-scale adoption of coke smelting by the British during the revolutionary period was a more effective damper to technological change in America, since the chronic shortage of iron in Great Britain was alleviated, thereby removing the single largest export market for surplus American production. Furthermore, a parliamentary act closing the ports of the West Indies to U.S. merchant ships eliminated the only other market to which iron in any quantity had been shipped during the colonial period.[55]

There remains the question of whether sufficient economic incentive existed in the substitution of coke for charcoal in the smelting process to justify following England's lead in that area. Unlike England, America possessed abundant forest resources suitable for charcoal smelting. Moreover, a characteristic form of American iron smelting unit, the iron plantation, had been organized around the need to ensure an adequate supply of wood. The relative cheapness of fuel close at hand fostered the maintenance of a charcoal iron industry in America long after English iron manufacture had been transformed by the use of coal and coke. Even when the American iron industry crossed the Appalachians into western Pennsylvania and Ohio, a region of abundant coal seams suitable for conversion to coke, the plantation model with its emphasis on charcoal smelting and forging remained the principal mode of operation well into the nineteenth century.[56] Considering the long history of iron manufacture in Lebanon County, Pennsylvania, Henry C. Grittinger observed, "Prior to 1836 charcoal was the only fuel used in the manufacture of iron in eastern Pennsylvania, although coke was used to a limited extent, in addition to raw coal, in the bituminous region."[57]

Another consideration—metal quality—may have been more important than relative fuel costs. Before 1780, it was generally agreed in England that the substitution of coke for charcoal in iron smelting resulted in an inferior product when converted to wrought iron. Coke smelting would have introduced deleterious impurities, primarily sulfur. A marked improvement in quality was achieved after 1784 by treating coke-cast iron by Henry Cort's patented puddling process, remelting the iron in a reverberatory furnace that burned off a large percentage of the impurities. The resulting pasty mass then was passed through grooved rolls, a process also patented by Cort, which squeezed out much of the higher melting impurities.[58]

In America, wrought iron for tools and hardware was a primary product of the iron industry, and metal quality was an important consideration, but sheer volume was not. Coal and coke may have been used to a limited extent in the heating and forging of iron in the eighteenth century, where the short exposure to the fire did not allow absorption of impurities. As Peter Temin has shown, when coke smelting finally was adopted in America after 1830, it was in conjunction with the later "hot blast" development, a system for preheating furnace gases to give higher temperatures, which also resulted in lower retention of sulfur. The resulting higher quality made the product economically competitive with charcoal iron, justifying the higher initial costs for coke-smelting furnaces.[59]

Immediate economic incentive, then, was lacking to adopt coke smelting and to build larger furnaces that would expand the capacity of the iron industry along the Atlantic seaboard. The war, however, had introduced yet another factor tending to oppose technological change. This was an acceleration of the movement of population away from the "war zones" of the coast westward into the wilderness. In the process, the inexpensive, convenient transport network of bays and tidewater rivers was left behind. Across the Appalachians, land transport to and from the East was so difficult and costly as to preclude the shipment of iron. Whereas Americans along the Atlantic seaboard once had deemed it imperative to transfer the technology of iron manufacture across an ocean, others reacted to the changing circumstances by spreading the iron industry westward. The needs of the slowly growing population were met by the construction of a series of charcoal blast furnaces in the new areas. The technology was well known, consistent quality of wrought iron could reasonably be assured, and the cost of charcoal smelting was lower than that for larger coke furnaces. Moreover, high volumes were not necessary for the relatively localized markets.

Shortly after 1790, the first ironworks in the trans-Appalachian region were built in northeastern Tennessee, near the present site of Kingsport.[60] Also about 1790, the first works in the present state of Kentucky, on the Licking River near Owingsville, began operations.[61] During the same years, the first furnaces and forges were constructed in western Pennsylvania, the forerunners of the great Pittsburgh iron complex of the mid-nineteenth century.[62] Until such time as dramatic changes in the consumption of iron made necessary larger furnaces, the old technology was adequate as well as less expensive on the scale required. In the United States, the level of demand did not rise to impel change until the introduction of the railroads in the 1820s. The technological lag in the American iron industry, if indeed it can be so construed, was the product of rational political, social, and economic forces.

The 1790s were a period of sustained growth for the industry, but growth utilizing a technology already becoming obsolescent in Great Britain. Exact figures for the number of iron facilities in the United States are not available before the census of 1810, which itself was incomplete. By that time, however, the United States contained over 150 furnaces, 465 forges and bloomeries, 34 slitting mills, and more than 410 naileries, the great majority of them having been built before the end of the eighteenth century.[63] Practically all of the additional production of pig and bar iron was consumed by the domestic markets. Average yearly exports of pig and bar iron for the 1790s were less than two thousand tons and seven hundred tons, respectively, with pig iron exports falling sharply toward the end of the period, while imports remained nearly constant. The growth of the iron industry was just keeping pace with the growing demand of the nation throughout the decade.[64]

Iron, although the largest metal industry in America, was not the only one to feel the disruptive effects of the war and its economic aftermath. Although, as Hamilton observed, "The material is a national production of the Country," copper mining in the United States had all but ceased by the end of the Revolution. The principal copper-mining region in New Jersey had been in the path of the armies surging back and forth across the state, and all operations had been suspended. Moreover, ample supplies of low-priced copper and brass pig and sheet entered the American market from Great Britain in the eighties, discouraging mining. Those conditions persisted for some years as the relative stability of the price of imported copper sustained the lack of enthusiasm for mining up to the outbreak of hostilities between France and England in 1793.[65] Then a growing demand for ship sheathing in England, coupled with military requirements, created a rise in the price of copper from 25¢ per pound in August 1791 to 50¢ per pound by August 1798. With the intensification of the war, by 1798, Great Britain was forced to place limitations on the export of copper.[66]

The rise in prices inspired new attempts to rework some of the abandoned copper mines. The largest effort was the reopening of the Schuyler Copper Mine in New Jersey in 1793 by a company formed by Jacob Marke, Philip A. Schuyler, and Nicholas Roosevelt. Besides erecting a stamp mill, foundry, and machine shop, the company unsuccessfully applied to Congress for monopoly privileges to search for and work copper mines in the western territories of the United States. Even the higher prices brought about by shifting international fortunes were not sufficient to ensure the success of such ventures. By about 1800, no profit had been realized from what was, during the colonial period, America's most successful copper mine.[67] Consequently, few attempts were made to find new mines. A vein of copper discovered during the period in Bristol, Connecticut, was not actively worked before 1836.[68]

Faced with the strong competition from imported British goods, the practice of coppersmithing also was curtailed. The craftsmen were active among those submitting petitions to Congress, but they received only slight benefit from the policy of taxing the import of finished products. A policy of allowing the entry of raw materials duty free was established by mid-decade and carried over into the Tariff Act of 1789. Well into the nineteenth century, copper and brass manufacturing was dependent upon foreign imports of raw materials.

For the only other metal mined during the colonial period, lead, the same conditions largely prevailed. The mine in Virginia operated on lease by a series of proprietors continued to produce lead, much of which was sent to Richmond for manufacture of shot and sheet lead for roofing.[69] Otherwise, lead mining had all but ceased in the United States. Through northern ports, an average of nearly two million pounds per year of lead pig and sheet was imported during the 1790s, primarily from Bristol, England.[70] In addition, the United States began to import quantities of white lead, a pigment for paints.[71] The active production of lead in America did not resume before the start of the nineteenth century. The volume of imports, even of base lead, was further evidence of the strong desire for metals in America and was indicative of the importance metals were to have in the years to come.

At the end of the eighteenth century, the future role of metals technology in the United States already had been determined. To a casual observer on the scene, it would have been hard to discern that the uncertain progress then being made was in the direction of the industrial greatness of the next century. True, iron was being produced by furnaces and forges in all of the states and was beginning to cross the Appalachians with the growing tide of western settlers. The metals trades were widely practiced, although not yet restored to the health of the prerevolutionary years. But dependence upon foreign imports for raw materials continued. Iron, copper, and lead were the metals in immediate demand; they had already established their value to the nation. The United States could supply itself only with iron. The principle of protection for the encouragement of manufactures, metals prominent among them, was embodied in tariff legislation. Hamilton's *Report* had emphasized the importance of metals. In 1800, if a full appreciation of that importance was not yet established, future events would make it manifest. The pattern of the colonial period was being repeated. The country was growing, expanding westward, and to support that growth it had to have metals. The problem of raw materials might be solved. It was known that there was lead on the Mississippi and copper in the Northwest Territories. Long dormant, but not forgotten, was the hope that the unknown area might contain even greater metallic wealth than any yet discovered in North America.

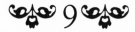

Reflections on the End of an Era

That Iron, as a metal, is of infinitely more real Use than Gold, is a Fact too obvious to stand in Need of Proof; since the very Necessaries of Life could neither be perceived, nor any useful Art or Manufactory carried on without the Assistance drawn from that valuable material.—Anon., England, 1757[1]

Throughout this study covering the period of the settlement and growth to independent statehood of the English-speaking colonies in North America, there has been one underlying hypothesis—metals were a central element in the material culture of Western civilization. Without fabricated metal artifacts, in particular, the successful colonization of America could not have been accomplished. Without continuing supplies of metals, both as raw materials and finished products, the growth of the colonies could not have been sustained. Without the transfer of metals technologies to the New World and the capacity to produce muskets and cannon to support Continental forces in the war against Great Britain, independence could not have been achieved.

If, however, metals played such an important role in the founding of the English-speaking colonies and in the subsequent shaping of their destiny, one might reflect on why, in general, so little attention has been devoted to metals in writing the history of colonial America.[2] Certainly the omission does not represent a failure to appreciate the importance of metal, particularly iron, in society. Iron and steel figure prominently in histories of the nineteenth- and twentieth-century United States. It seems more likely that the oversight has resulted from what might be termed the "low profile" projected by metals before 1800, a profile determined by the complex pattern of development of the technologies and colored by assumptions normally stemming from the role those technologies would play in any colonial society.

Throughout much of the colonial period, the settlements of the English-speaking peoples were, in fact, little more than outposts in the wilderness. The small population was scattered along thirteen hundred miles of coastline with few concentrations exceeding five thousand people, the size of many modern small towns. Along much of that coastline the undisturbed primeval forest came down to the sea, and elsewhere the westward frontier of settlement was rarely more than a day's journey inland. The tenuous maritime lifeline stretching from England would have appeared more than adequate to satisfy the material requirements of a population so situated. Besides lacking urban centers for markets or clearly defined centers for manufacturing and trade, there were, with the few exceptions of a Hammersmith or Principio Company, no deliberate, concerted efforts to transfer to America the equipment and skill associated with metals technology in England. The lack of well-defined movements of technology from England to America affords no hint of the importance of metals to the American colonies throughout the seventeenth and well into the eighteenth centuries. Furthermore, external clues that might lead the historian to the subject are lacking, and the internal documentary evidence concerning metals too often is scanty or misleading.

Only on a few occasions during the colonial period was an awareness of the importance of metals explicitly stated in the words of the colonists. In the initial stages of exploration and colonization, that awareness took the form of a clearly expressed search for gold and silver. As Sir Thomas More long before had perceived, however, gold and silver were the "wrong" metals in terms of their relative value. They were the wrong metals in terms not of desirability but of functional usefulness. Moreover, the search for them delayed the introduction of the basic technologies necessary to sustain and promote the growth of the colonies. From our perspective, they also were the wrong metals in the sense that the widely heralded search for them and the obvious failure to find them have acted to obscure the far greater importance of baser metals, particularly iron, to this period.

For a brief time in seventeenth-century Massachusetts, the General Court gave explicit recognition to the need for iron in the colony by granting a conditional monopoly to the Company of Undertakers of the Iron Works in New England to build and operate furnaces at Braintree and Saugus. The monopoly was to be effective for twenty-one years provided that "within two years they make sufficient iron for the use of the country." When the company no longer was able to comply with the court's command, the monopoly was revoked, but any overt consideration of the colony's need for iron was submerged in the flood of litigation that marked the company's end.[3] It did not reappear anywhere in colonial America in the seventeenth century.

The general lack of explicit expression of the need for metals throughout most of the colonial period, was not, however, an accurate reflection of the actions of the colonists to whom the need for metals was implicit. By the eighteenth century, a pattern of development of metals technologies had begun to emerge along well-defined geographical lines.

Yet in the eighteenth century, the occasional actions of colonial legislatures in granting patents to erect ironworks or in establishing monopolies for metals manufactures were obscured by the broader issues of colonial regulation that began to occupy authorities on both sides of the Atlantic.

To unravel the interplay of complex factors that in the eighteenth century culminated in the creation of an indigenous iron industry in America, a utopian analogy has proved useful in the present study. When the analogy was applied to the growth of iron technology, a sequential evolution was found that, by extension, also encompassed colonial activity in the mining of copper. The development of metals technologies can be grouped into three distinct stages. The first I have called the prospecting stage, where the main object was to find gold or silver, and possibly copper, that could yield quick and substantial returns.

The second or entrepreneurial stage, as I have defined it, was an attempt to exploit the metal ores that could be found in America, principally iron and copper, with profits realized from the sale of ore or refined metal in Great Britain. It was in the entrepreneurial spirit that the first furnaces were built at Falling Creek, Virginia (1622), and at Saugus (1646). The same impetus later contributed to the creation of the iron complexes in the Chesapeake area, Spotswood's and the Principio and Baltimore companies. It also was a factor in the erection of some of the early Pennsylvania furnaces.

The final stage was the creation of an indigenous industry, the building of furnaces and forges to refine and manufacture iron to meet the needs of the growing colonial population. The beginnings of this indigenous phase have been ascribed to the works erected in eastern Pennsylvania early in the eighteenth century. Moreover, because of its central importance to material culture, only the smelting and manufacture of iron reached the indigenous stage during the colonial period.

The transfer of the technologies of mining and refining that accompanied the evolution of the general pattern had a corresponding characteristic pattern. As considered in Chapter 2, the English people apparently possessed little skill in the identification of minerals. That inherent weakness contributed to the relative failure of the early colonists to realize their ambitions, at least those who had visions of discovering precious metals. The eastern seaboard of North America did have limited mineral wealth, but the difficulty experienced in locating any ores other than iron or an occasional deposit of copper must be attributed to the lack of prospecting know-how among the early colonists. There is no evidence to suggest that skill in discovering and identifying minerals of value improved during the entire colonial period.

The one metal whose ores the English could identify was iron. To exploit the obviously widespread deposits of iron ore in America two alternatives were available: one was to ship the ore to England for refining, as was initially done at Jamestown; the second was to build the furnaces in America. The latter was clearly the more attractive, because of abundant availability of wood for smelting the ore and the relatively low yield of the ore itself. A consistent average yield of 10 percent iron by weight would have been high, meaning that 90 percent of the cost of transporting the ore would have been for useless dross. (Compare this with the case of high-grade copper ores, such as those initially mined in Connecticut and New Jersey, where a 50 percent yield would not have been unusual.)[4]

To build furnaces, workers experienced in the different phases of the refining, casting, and manufacture of iron were sent to America, thus effecting the transfer of the technology. The Virginia Company of London, the Company of Undertakers, and the Principio Company all went to great pains to procure skilled workmen in England for the American enterprises. Colonel Spotswood in Virginia and Peter Hasenclever in New Jersey resorted to the expedient of importing large numbers of German workmen. In one case of copper smelting at Simsbury, Connecticut, Hanoverian artisans apparently were imported to construct the works. Unlike the prospecting stage, the entrepreneurial stage carried its technology with it.

The introduction of the metals crafts in the seventeenth century may be considered a precursor of the indigenous stage since, as shown in Chapter 5, the great increase in their practice in the mid-eighteenth century coincided with the most rapid growth of the iron industry. From the very beginning of settlement, the introduction of the metals crafts began to occur. The presence of goldsmiths in Jamestown in 1607 may be attributed to an excess of optimism on the part of the London Company. The presence of James Read, blacksmith, can only be ascribed to practical necessity, the need for the maintenance of a broad range of tools, agricultural, and mining implements.[5]

The pattern at Jamestown was repeated elsewhere, blacksmithing being among the first of the skilled trades to be practiced in settled areas. The earliest introduction of the metals crafts occurred while the colonies were still totally dependent upon imports from Great Britain for raw materials as well as the greatest part of their supply of manufactured goods. Except for the attempts to find and mine gold and silver and the one abortive attempt to mine and refine iron at Falling Creek, the crafts were the first metals technologies transferred to America. Although the techniques involved in shaping metals may, in themselves, have been sophisticated, they were also among the easiest forms of technology to bring across the Atlantic. The technology of the metal artisan consisted of two elements—knowledge acquired by the craftsman through a protracted apprenticeship and therefore transferred by his presence, and tools. We know very little about the transfer of artisan's tools except that they, like much of the raw material, had to be imported during most of the colonial period.[6]

The beginnings of the indigenous stage followed a different line of development. A certain amount of empirical knowledge already was present as a result of previous entrepreneurial endeavors; later New

England ironworks benefited from the reservoir of skills established at Saugus. But the men who built the first furnaces in the forests of eastern Pennsylvania and who constructed many of the early ironworks in New York, New Jersey, and Maryland had not come to America to participate in any of the exploitative ventures. Their skills were from the beginning devoted to serving a local market, first at the forge and then, as demands upon their skills increased from a growing population and rising prosperity, in the building and operation of furnaces to produce iron for America. As ironworks grew, they tended to draw heavily upon the skills of the craftsmen already in the colonies. By ensuring adequate supplies of cast and wrought iron, they in turn helped to further the growth and spread of the metals crafts.

The progression from the presence of blacksmiths using imported or recycled stock in the early years of colonization to the creation of an indigenous iron industry that, by the start of the Revolution, rivaled that of Great Britain was the result of developments on both sides of the Atlantic. Two factors were necessary for its creation. The first was a rising degree of prosperity to provide the funds necessary for transforming the technology of iron smelting from simple bloomery forges to the integrated blast-furnace complexes. The second was the existence of local markets to absorb the bulk of the furnace output. The relatively lax mercantile policies of England, promoting economic growth in the colonies, helped America to acquire the necessary capital. The steady expansion of the colonial population with the consequent rise of cities and internal trade networks provided the market. But it was the inability of the British iron industry to expand rapidly enough to meet growing demands at home and abroad that made the creation of that indigenous industry inevitable.

The three stages of development of metals technologies did not follow one upon the other in clear succession but represented shifting priorities. In that sense, the pattern resembled overlapping waves with distinct crests, the last of which, corresponding to the indigenous stage, was rapidly growing at the end of the period. Thus evidence of the prospecting stage could be discerned in Alexander Henry's attempt to mine copper ore for its silver content in the Lake Superior region as late as the 1760s. The operation of the Principio Company throughout the eighteenth century was an extension of the entrepreneurial stage. But the failure of Peter Hasenclever's American Iron Company, founded by British interests to produce iron for a British market, symbolized the shift in emphasis from the entrepreneurial, exploitative stage to the indigenous, domestically oriented phase on the eve of the Revolution. Hasenclever built well, but his operation was anachronistic. The market for American iron no longer was eastward across the Atlantic but in the rising cities of the colonial seaboard and in the needs of a population moving westward toward the Ohio. While Hasenclever looked to England, the tide of history was flowing past him, building in the opposite direction.

The establishment and expansion of an indigenous American iron industry called the attention of Parliament to the growth of manufactures in the British colonies. Even more important than the competitive threat such manufactures posed to British iron interests were the long-range implications of allowing those manufactures to continue and to prosper. Although parliamentary debates focused on the economic effects, early in the eighteenth century Lord Cornbury, Sir John Trenchard, and a few other perceptive individuals recognized that the establishment of manufactures in the colonies was, potentially, a first step toward ultimate independence. By midcentury, resistance to the regulation of iron manufactures leading to open defiance of such regulation justified their warnings. A movement toward that independence was beginning, and the manufacture of metals was an integral part of that movement.

Once more, however, other issues obscured the role of metals in reaching the break. In this case it was the swelling chorus that argued the justification for dissolving the political bonds that connected one nation to another. That the metals technologies long since established and flourishing were an important factor in the conduct of the War for Independence was overshadowed in the broader perspective of international crises. Finally, the aftermath of the war saw the initiation of the great experiment in government leading to the Constitution of the United States. Concern with the problems of government and the survival of the new nation dominated the speeches and writings of leading figures. Throughout the critical years little overt attention was given to metals or other manufactures or to their future in the new nation.

One final factor contributing to the relative obscurity of metals in the colonial period merits consideration. By definition, colonial status implies a degree of dependency, and one criterion for measuring the degree of dependency is an analysis of imports and exports. Throughout the colonial period, American imports from Great Britain contained a large percentage of manufactured goods, while exports from the colonies were predominantly those to be expected of an almost totally agrarian society. To cite a single example, during 1770, exports of pig and bar iron to Great Britain accounted for only 6 percent of the total trade valued at £1.5 million sterling.[7] During the same year, however, by considering the number of active forges and furnaces within the colonies, it can be estimated that the production of twenty thousand tons each of pig and bar iron would have had an approximate market value of £300,000. Hence a clear impression of the relative size and importance of the iron industry cannot be derived from examining the export data. The absence of comparable analyses of intracolonial trade has obscured the economic significance of iron to colonial commerce.

Again for 1770, it can be estimated that the total number of men employed in the primary stages of iron production, the actual reduction of iron from ore at a furnace or bloomery forge, probably did not exceed eight thousand out of a population of two million. This means that, in percentage terms, only about 0.4 percent of the population was required to supply the basic needs in iron for the entire country in 1770.[8] At face value, this appears to be one of the strongest possible arguments for downplaying the significance of the iron industry and, by extension, since it was largest, all metal manufactures in the colonial period.

Consider, however, what would happen were we to evaluate the modern iron industry by the same criteria. Today, the importance of iron is unquestioned. In 1970, enormous blast furnaces in conspicuous concentrations in places like Pittsburgh, Gary, and Birmingham spewed out over 90 million tons of pig iron and another 120 million tons of carbon steel (a composition roughly corresponding to colonial bar iron). A population one hundred times greater than that of colonial America used over one ton of iron and steel per person for 1970, representing a ten-thousandfold increase in the capacity of the iron and steel industry over the intervening two centuries. The great increase in production, as well as consumption, was made possible by sophisticated changes in iron technology. Yet, in 1970, the entire primary production of iron to serve the needs of the modern United States required the labor of only 0.4 percent of the population.[9] The difference in output between 1770 and 1970 is one of scale. The technological functions of modern society are clearly dependent on the production of iron and steel and a host of other metals. The lower production figure for 1770, however, should not be allowed to obscure the fundamental importance of iron, or of copper, lead, and tin to that earlier era.

Alexander Hamilton, perhaps of all Americans of the colonial period, came closest to expressing the true significance of metals for America when he advocated the protection and encouragement of the metal technologies in the *Report on Manufactures*. The importance of metals rested not in numbers nor was it meant to be synonymous with industrialization. It rested on the need to fulfill the basic material requirements of American society. To allow the United States to become dependent upon another nation for those material requirements could mean the slow strangulation and destruction of the civilization and the nation. The *Report on Manufactures* was a last clear articulation of the role metals had played in the growth of the colonies and the founding of the nation. It was both a recognition and a reminder that, as the needs of society continued and expanded, so must the means to satisfy those needs continue and expand. The manufacture of metals had a role to play in the future of the United States. For Hamilton, the Revolution had marked the end of political colonialism in America. It was time for the colonial era in metals manufactures to end also.

If any one event can be considered as symbolizing the end of the colonial era in American metals, perhaps it was the discovery of gold in western North Carolina in 1799. Tradition relates how Conrad Reed, a young farm boy, noticed a colorful rock in a stream running through the family property. The boy took it home, where it gathered dust as a doorstop for three years. Finally, his father took the rock, later estimated to be a pure gold nugget weighing seventeen pounds, to Fayetteville where he sold it to a jeweler for $3.50.

Conrad Reed's chance find had answered an old question. The precious metals so eagerly sought by the early explorers and colonists did in fact exist but, ironically, were not immediately recognized when finally located. Twenty years later, the eastern slope of the Appalachians from Virginia to Georgia became the scene of the first gold rush in the United States. Reed's discovery, however, was an anticlimax.[10] Long before, the colonists had discovered that the need for the manufacture of metals was more important than the presence or absence of gold. Consequently, they developed the technologies to mine and smelt ores and to fabricate artifacts from several metals. In the process, they helped to transfer the full dimension of Western civilization to America. Moreover, the manufacture of metals aided the growth of the colonies and the formation of a new nation. The colonists' discovery and their efforts to establish the metals technologies proved to be of greater value to America than all the gold of the Indies.

Notes

The following sources are cited frequently throughout the footnotes. For simplicity, a shorthand form of citation is used, as indicated below.

Great Britain, Public Record Office, *Calendar of State Papers, Colonial Series, Preserved in the Public Record Office*, 38 vols. (London, 1860–1919). Hereafter, *Calendar of State Papers, Colonial*.

Worthington Chauncey Ford et al., eds., *Journals of the Continental Congress, 1774–89*, 34 vols. (Washington, D.C.: 1904–37). Hereafter, *Journals of Congress*.

William Bell Clark and William James Morgan, eds., *Naval Documents of the American Revolution*, 7 vols. to date (Washington, D.C.: 1964–). Hereafter, *Naval Documents*.

J. Hammond Trumbull and Charles Jeremy Hoadley, comps., *Public Records of the Colony of Connecticut, 1636–1776*, 15 vols. (Hartford, 1850–90). Hereafter, *Conn. Records*.

William Hand Browne et al., eds., *Archives of Maryland*, 70 vols. (Baltimore: 1883–1964). Hereafter, *Md. Archives*.

Nathaniel B. Shurtleff, ed., *Records of the Governor and Company of the Massachusetts Bay in New England, 1628–86*, 5 vols. (Boston, 1853–54). Hereafter, *Mass. Records*.

William Adee Whitehead et al., eds., *Archives of the State of New Jersey*. 1st series, *Documents Relating to the Colonial History, 1631–1800*, 42 vols. (Newark, etc., 1880–1949). Hereafter, *N.J. Archives*.

Edmund Bailey O'Callaghan and Berthold Fernow, eds., *Documents Relative to the Colonial History of the State of New York*, 15 vols. (Albany, 1856–87). Hereafter, *N.Y. Records*.

William L. Saunders and Walter Clark, eds., *Colonial and State Records of North Carolina, 1662–1790*, 26 vols. (Raleigh, Winston, etc., 1886–1906). Hereafter, *N.C. Records*.

Samuel Hazard, et al., eds, *Pennsylvania Colonial Records, 1683–1790*, 16 vols., vols. 1–10; *Minutes of the Provincial Council of Pennsylvania from the Organization to the Termination of the Proprietary Government*; vols. 11–16; *Minutes of the Supreme Executive Council* (Philadelphia and Harrisburg, 1838–60). Hereafter, *Penn. Records*.

Samuel Hazard, ed., *Pennsylvania Archives*, 1st series, 12 vols. (Philadelphia: 1852–56). Hereafter, *Penn. Archives*.

John Russell Bartlett, ed., *Records of the Colony of Rhode Island and Providence Plantation in New England, 1636–1792*, 10 vols. (Providence, 1856–65). Hereafter, *R.I. Records*.

Harold C. Syrett, ed., *The Papers of Alexander Hamilton*, 26 vols. to date (New York: Columbia University Press, 1961–). Hereafter, Hamilton, *Papers*.

Danby Pickering et al., comps., *Statutes at Large from Magna Charta to 1807*, 46 vols. (Cambridge and London, J. Bentham, 1762–1807). Hereafter, Pickering, *Statutes*.

Introduction

1. Quoted in Carl L. Becker, *The Heavenly City of the Eighteenth-Century Philosophers* (New Haven: Yale University Press, 1932), p. 88.

2. Sir Thomas More, *Utopia*, H.V.S. Ogden, trans. (New York: Appleton-Century-Croft, 1949), p. 43.

3. Ibid.

4. The role of metal technologies and European naval dominance is well established in Carlo M. Cipolla, *Guns, Sails, and Empires* (New York: Pantheon Books, 1965).

5. Joseph E. Walker, "The End of Colonialism in the Middle Atlantic Iron Industry," *Pennsylvania History*, 41 (1974):5–26.

6. Alexander Hamilton, *Report on Manufactures*, in Hamilton, *Papers*, 10:230–340.

Chapter 1

1. More, *Utopia*, p. 43.

2. George Chapman, "Eastward Ho," act 3, scene 3, in *The Plays and Poems of George Chapman: The Comedies*, Thomas Marc Parrott, ed. (London: Russell and Russell, 1914), p. 498. Although the play was jointly written by Chapman, Ben Jonson, and Joseph Marston, Parrott attributes this particular scene to Chapman. See his analysis, p. 845.

3. More, *Utopia*, pp. 27, 42–43.

4. John U. Nef, "Mining and Metallurgy in Medieval Society," in M. Poston et al., eds., *The Cambridge Economic History of Europe*, 6 vols. (Cambridge: Cambridge University Press, 1941–65), 2:469–71.

5. See Cyril Stanley Smith, "The Discovery of Carbon in Steel," *Technology and Culture*, 5 (1964):149–75.

6. Dudley T. Easby, Jr., "Pre-hispanic Metallurgy and Metalworking in the New World," *Proceedings of the American Philosophical Society*, 109 (1963):89–98, considers the techniques of both the Aztec and Incan civilizations by examining surviving art works.

7. Ibid., pp. 90–94; also Leslie Aitcheson, *A History of Metals*, 2 vols. (London: Macdonald & Evans, 1960), 2:363–67.

8. Henry R. Wagner, "Early Silver Mining in New Spain," *Revista de historia de America*, 14 (1942):49–71, discusses Spanish efforts to locate and mine silver. Verifiable accounts of mining before 1530 are rare.

9. Neville Williams, *The Life and Times of Henry VII* (London: Weidenfield and Nicolson, 1973), p. 144.

10. "English Letters patent to John Cabot, 1496," *English Colonization of North America*, Louis B. Wright, Elaine W. Fowler, eds. (London: Edward Arnold, 1968), pp. 11–12.

11. Samuel Eliot Morison, *The European Discovery of America: The Northern Voyages, A.D. 500–1600* (New York: Oxford University Press, 1971), pp. 157–209, summarizes the scholarship on the subject. A short, annotated Cabot bibliography is included, pp. 192–93.

12. James Alexander Williamson, "England and the Opening of the Atlantic," in J.H. Rose et al., eds., *The Cambridge History of the British Empire*, 9 vols. (Cambridge: The University Press, 1929–59), 1:25–28.

13. By the time of his death, it was estimated that Henry's coffers contained at least £1,300,000 in gold, none of it from Cabot's voyages, however. See

Frederick C. Dietz, *English Government Finance: 1485–1550* (1921; reprint ed., New York: Johnson, 1970), pp. 78–87, for the value of Henry's revenues.

14. John U. Nef, *War and Human Progress* (Cambridge: Harvard University Press, 1950), pp. 24–25.

15. Aitcheson, *History of Metals,* 2:390–93. A more detailed analysis of the changes in British iron technology is given by H. R. Schubert, *History of the British Iron and Steel Industry from c. 450 B.C. to A.D. 1775* (London: Routledge & Kegan Paul, 1957), pp. 161–72.

16. Maxwell Bruce Donald, *Elizabethan Copper* (London: Pergamon Press, 1955), p. 12.

17. Richard Hakluyt, *The Principal Navigations, Voyages, Traffiques and Discoveries of the English Nation,* 10 vols. (New York: E. P. Dutton, 1927–28), 1:212–17.

18. Morison, *Northern Voyages,* pp. 285–87, 305.

19. Pope Clement was an ally of the king of France against the Spanish emperor, Charles V, at the time this edict was pronounced.

20. James Phinney Baxter, *A Memoir of Jacques Cartier, His Voyages to the St. Lawrence* (New York: Dodd, Mead, 1906), pp. 75 ff., 189–90; Joseph E. King, "The Glorious Kingdom of Saguenay," *Canadian Historical Review,* 31 (1950):390–400.

21. Louis B. Wright, *The Dream of Prosperity in Colonial America* (New York: New York University Press, 1965), pp. 6–12; Baxter, *Jacques Cartier,* pp. 339–40.

22. Morison, *Northern Voyages,* pp. 433–37; Baxter, *Jacques Cartier,* pp. 335–38.

23. Hakluyt, *Principal Navigations,* 9:444.

24. Morison, *Northern Voyages,* pp. 430–54.

25. John Huxtable Elliot, *Europe Divided: 1449–1598* (London: William Collins & Son, 1968), pp. 166–67.

26. William Camden, *A History of the Most Renowned and Victorious Princess Elizabeth, Later Queen of England,* 4th ed. (London: M. Flesher, 1683), pp. 103–05.

27. Charles Gibson, *Spain in America* (New York: Harper & Row, 1966), pp. 103–05.

28. Anne Fremantle, ed., *The Papal Encyclicals in Their Historical Context* (New York: New American Library, 1956), pp. 82–85.

29. Donald, *Elizabethan Copper,* Appendix II, contains a list of more than one hundred Germans associated with the English copper mines and their occupations. The preponderance were miners, smelters, or coppersmiths (pp. 373–76).

30. Camden, *Princess Elizabeth,* p. 215.

31. The black stone brought back from the first voyage is mentioned by nearly all writers discussing Frobisher's expeditions. An extensive account and analysis is in Vilhjalmur Stefansson, *The Three Voyages of Martin Frobisher,* 2 vols. (London: Argonaut Press, 1938), which includes the texts of all existing contemporary accounts. The ore incident and its ramifications are documented in 2:81–94. See also Morison, *Northern Voyages,* pp. 508–10, 516–17.

32. Stefansson, *Martin Frobisher,* 1:cxi–cxiii.

33. Morison, *Northern Voyages,* p. 531.

34. *Calendar of State Papers, Colonial,* Eastern Series, *East Indies, China and Japan; 1513–1616,* 2:89.

35. Camden, *Princess Elizabeth,* p. 216; Stefansson, *Martin Frobisher,* 1:cxvi.

An attempt has been made to determine the exact composition of Frobisher's "ores" by Thomas Arthur Rickard, and his results are presented by Stefansson in Appendix 10, 2:248–52. In addition to pyrites, the rocks probably contained some horneblende and traces of mica, but no gold or silver.

36. Cyril Stanley Smith and Martha Teach Gnudi, trans., *The Pirotechnia of Vannoccio Biringuccio* (New York: The American Institute of Mining and Metallurgical Engineers, 1942), p. 135.

37. Ibid., pp. 458–61.

38. Georgius Agricola, *De Re Metallica*, Herbert Clark Hoover and Lou Henry Hoover, trans. (London: The Mining Magazine, 1912), pp. 599–603, 608–15.

39. Camden, *Princess Elizabeth*, pp. 254 ff.

40. For a historical sketch of the Norumbega myth, see Morison, *Northern Voyages*, pp. 464–70, Bibliography, pp. 488–89.

41. Ingram's testimonies are recorded in David Beers Quinn, ed., *The Voyages and Colonizing Enterprizes of Sir Humphrey Gilbert*, 2d series, vols. 83, 84 (London: Hakluyt Society, 1940), 84:281–88. Richard Hakluyt, *The Principal Navigations, etc.*, Facsimile Edition, D. B. Quinn, R. A. Shelton, eds., 2 vols. (London: Hakluyt Society, 1965), 2:557–59.

42. Quinn, ed., *Gilbert*, 84:313–35, 396.

43. Ibid., 84:412–14, 420–21. Quinn suggests that Gilbert's book was *Utopia* by comparing what Edward Hayes related were the captain's last words, "We are as near to heaven by sea as by land," to a passage from More, "The way to heaven out of all places is of like length and distance" (83:89, 89n).

44. E. G. R. Taylor, ed., *The Original Writings and Correspondence of the Two Richard Hakluyts*, 2 vols. (London: Hakluyt Society, 1935), 2:248–49.

45. Alexander Brown, *The Genesis of the United States*, 2 vols. (Boston: Houghton Mifflin, 1891), 1:20.

46. Hakluyt, *Principal Navigations*, 6:171.

Chapter 2

1. Michael Drayton, "Ode to the Virginia Voyage," *The Oxford Book of English Verse* (Oxford: Clarendon Press, 1926), p. 171.

2. Brown, *Genesis*, 1:52–53, 57–58. It may be indicative of the state of the English mind that the only metals specifically mentioned in the patents of that time were gold, silver, and copper, despite the frequent notice of bountiful supplies of iron. No provision was made for the finding of other mines or minerals or the exploitation thereof.

3. Edward Duffield Neill, *History of the Virginia Company of London* (Albany: J. Munsell, 1869), pp. 8–14.

4. Philip Alexander Bruce, *Economic History of Virginia in the Seventeenth Century*, 2 vols. (New York: The Macmillan Co., 1907), 1:21–41.

5. Wesley Frank Craven has examined the history of the Virginia Company and argues that the early seeking after gold at Jamestown, though one of the hopes on which the company was founded, has been greatly overemphasized by historians. See his *Dissolution of the Virginia Company* (Gloucester: Peter Smith, 1964), pp. 24–29.

6. John Smith, "A Map of Virginia" (1612) in Edward Arber and A. G. Bradley, eds., *The Travels and Works of Captain John Smith*, 2 vols. (Edinburgh: J. Grant, 1910), 1:49; italics in original.

7. *The Royal Commission on Historical Manuscripts: The Third Report* (London, 1872), p. 54.

8. Brown, *Genesis*, 1:111–13.

9. Thomas Studley and Anas Todkill, "The Proceedings of the English Colony in Virginia," in Arber and Bradley, eds., *John Smith*, 1:104.

10. Neill, *Virginia Company*, p. 22.

11. Arber and Bradley, eds., *John Smith*, 1:61.

12. Bruce, *Economic History*, 1:17–18

13. Arber and Bradley, eds., *John Smith*, 1:171–72.

14. *Calendar of State Papers, Colonial*, Eastern Series, *East Indies, China and Japan; 1513–1616*, 2:181. The specific reference is in the Court Minutes for February 13, 1609: "4 l. a ton to be paid for 17 tons of cider [sic, sidere (fr), iron] belonging to the Virginia Company."

15. Arber and Bradley, eds., *John Smith*, 1:12–13.

16. Brown, *Genesis*, 1:356.

17. Neill, *Virginia Company*, p. 43.

18. Brown, *Genesis*, 1:398.

19. John L. Cotter, *Archaeological Excavations at Jamestown*, National Park Series Archaeological Research Series No. 4 (Washington, D.C.: U.S. Department of the Interior, 1958), pp. 11, 110–13. An extensive discussion of the manufacture of iron in the Jamestown colony is the subject of Charles E. Hatch, Jr., and Thurlow Gates Gregory, "The First American Blast Furnace, 1619–1622," *The Virginia Magazine of History and Biography*, 70 (1962): 259–96.

20. William Strachey, *Historie of Travell into Virginia Britannia* (London: Hakluyt Society, 2d series, 1953), 103:131–32. See chapter 3 for an account of the attempts to discover and work copper deposits in North America.

21. Wesley Frank Craven, *A History of the South: The Southern Colonies in the Seventeenth Century, 1607–1689* (Baton Rouge: Louisiana State University Press, 1949), pp. 94–95.

22. Ibid., pp. 108–09, 135.

23. Schubert, *British Iron*, pp. 112–14.

24. "Nova Britannia: Offering Most Excellent fruites by Planting in Virginia" (1609), Peter Force, ed., *Tracts and Other Papers, Relating Principally to the Origin, Settlement, and Progress of the Colonies in North America*, 4 vols. (Washington, D.C.: P. Force, W. Q. Force, 1836–46), 1, no.4:16

25. "A true declaration of the estate in Virginia," ibid., p. 25.

26. A full account of the incident is given in Susan Myra Kingsbury, ed., *The Records of the Virginia Company of London*, 4 vols. (Washington, D.C.: U.S. Government Printing Office, 1906–35), 1:585–88.

27. Ibid., 3:116, 309.

28. Ibid., 1:472.

29. "Relation of Waterhouse," in Neill, *Virginia Company*, p. 338.

30. Hatch and Gregory, "First American Blast Furnace," pp. 284–91.

31. Neill, *Virginia Company*, p. 329.

32. Kingsbury, *Records of the Virginia Company*, 4:176–78.

33. "The King to the Governor and Council of Virginia" (Abstract) *Colonial Papers*, vol. 4, no. 32, quoted in *The Virginia Magazine of History and Biography*, 16 (1908): 34–35.

34. "A Description of New-England," in Arber and Bradley, eds., *John Smith*, 1:187, 201.

35. Bernard Bailyn has explored the role of trade in the colonization and growth of New England in his book, *The New England Merchants in the Seventeenth Century* (Cambridge: Harvard University Press, 1955). His well-balanced account shows how shifting fortunes in trade led to the codification of interests opposed to the basic religious motivation of the Puritan fathers. The development of an iron industry in Massachusetts is portrayed as a logical phase of the changing patterns of trade and politics.

36. A sensitive description of the background to the formation of the Massachusetts Bay Colony may be found in Edmund S. Morgan, *The Puritan Dilemma* (Boston: Little, Brown, 1958), pp. 18–44.

37. *Mass. Records*, 1:23. The units quoted represent: ffagot, 120 pounds, but signifying a bundle of wrought iron or steel that subsequently would be worked into a finished form by hammering while hot; ffoder (or fother), a variable unit of weight for lead, but on the order of one ton.

38. Ibid., 1:25–26, 36.

39. Higginson's letter is cited in Henry Wykoff Belknap, *Trades and Tradesmen of Essex County, Massachusetts* (Salem: The Essex Institute, 1929), pp. 3–4.

40. Ibid., p.14.

41. *Mass. Records*, 1:28, 30, 32–33.

42. Bailyn, *New England Merchants*, pp. 45–74, which includes his interpretation of the Saugus iron venture.

43. *Mass. Records*, 1:327.

44. For Winthrop's role in Massachusetts iron manufacture, see Bailyn, *New England Merchants*, pp. 62–67; Robert C. Black, III, *The Younger John Winthrop* (New York: Columbia University Press, 1966), pp. 110–32; and Edward Neal Hartley, *Ironworks on the Saugus* (Norman: University of Oklahoma Press, 1957), pp. 44–116, which present differing interpretations of the significance of Winthrop's participation. The activities of the mission are described by Raymond Phineas Stearn, "The Weld-Peter Mission to England," *Publications, Colonial Society of Massachusetts*, 32 (1931):188–246.

45. Winthrop impressed English as well as colonial contemporaries. He was the first colonist to be elected a member of the Royal Society. His projects were legion, and few of them were ever completed. Few occupied as much of his attention as the building of ironworks.

46. Charles A. Browne, "Scientific Notes from the Books and Letters of John Winthrop, Jr.," *Isis*, 9 (1928):325–41.

47. Black, *John Winthrop*, pp. 117–20; *Winthrop Papers*, 6 vols. (Boston: The Massachusetts Historical Society, 1929–42), 4:424–25. Winthrop was not able to obtain a bloomery man among his nucleus of skilled workmen. The need to import expertise in all phases of metals manufacture continued throughout most of the colonial period.

48. *Winthrop Papers*, 4:425–26.

49. *Mass. Records*, 2:61–62.

50. Hartley, *Saugus*, pp. 90–99.

51. Ibid., pp. 103–10.

52. *Winthrop Papers*, 5:140; Hartley, *Saugus*, pp. 122–23.

53. Cf. Bailyn, *New England Merchants*, pp. 65–66; Black, *John Winthrop*, pp. 122–25; *Winthrop Papers*, 4:463–64. Winthrop's iron activities in the New Haven Colony are discussed more fully in Chapter 4.

54. Hartley, *Saugus*, pp. 4–5.

55. The physical description of the original Hammersmith plant follows that given by Hartley, ibid., pp. 161–81.

56. The figure of four hundred bushels of charcoal for one ton of pig iron is approximate. Arthur Cecil Bining noted a rate of consumption for Oley Furnace in 1783 at "more than 400 bushels" per ton (*Pennsylvania Iron Manufacture in the Eighteenth Century* [Harrisburg: Pennsylvania Historical and Museum Commission, 1938], p. 63). In contrast, Hartley's figure of 265 bushels of charcoal per ton of pig iron (*Saugus*, p. 174), appears low.

57. An example of the stability of technology is the fact that modern, high-speed slitting mills, used in the processing of nearly all metals, are identical in form and principle of operation to the mill at Hammersmith.

58. *Winthrop Papers*, 5:246, 248.

59. George Francis Dow, ed., *Records and Files of the Quarterly Courts of Essex County, Massachusetts*, 8 vols. (Salem: The Essex Institute, 1911–21), 2:74–98. The June 1658 session contains a long series of writs and depositions pointing to the disorganized state of Gifford's financial transactions. The facts came to light after Gifford had been reinstated in 1657.

60. Hartley, *Saugus*, pp. 162–63, 215–43.

61. Ibid., pp. 272–305. See also discussion in Chapter 4.

62. Ibid., p. 80.

Chapter 3

1. "Gov. John Winthrop to John Winthrop, Jr., September 30, 1648," *Winthrop Papers*, 5:262

2. *N.Y. Records*, 1:122–23.

3. Ibid., 1:148.

4. Letters of August 31, 1645, and December 1646 in West India Company correspondence, quoted by Charles Gilbert Hine in *The Old Mine Road* (New Brunswick: Rutgers University Press, 1963, reprint of *Hine's Manual* for 1908), p. 2.

5. *N.Y. Records*, 1:262.

6. Ibid., 1:280

7. Ibid., 2:63.

8. See Hine, *Mine Road*, pp. 1–21, 159–61; John Cary, "Historic Resources of the Upper Delaware: The Jersey Side," *Proceedings of the New Jersey Historical Society*, 83 (1965):84–85; Harry B. Weiss and Grace M. Weiss, *Old Copper Mines of New Jersey* (Trenton: Past Times Press, 1963), pp. 82–85.

9. Clarence Walworth Alvord and Lee Bidgood, *The First Explorations of the Trans-Allegheny Region by the Virginians, 1650–1674* (Cleveland: A. H. Clark, 1912), pp. 101–02.

10. Ibid., pp. 93, 102–04.

11. Ibid., pp. 61–62, 162, 171.

12. Gold was initially discovered in North Carolina in 1799. That discovery and the prior search are summarized in Fletcher Melvin Green, "Gold Mining, A Forgotten Industry of Ante-Bellum North Carolina," *The North Carolina Historical Review*, 14 (1937):3–8.

13. The native copper of the Lake Superior regions was nearly 99.9 percent pure, and small quantities of native silver were found with it also.

14. George Brinton Phillips, "The Primitive Copper Industry of America," Part I, *Journal of the Institute of Metals*, 34 (1925):261–67, and Part II, ibid., 36 (1926):99–106.

15. Roy Ward Drier and Octave Joseph Du Temple, eds., *Prehistoric Copper Mining in the Lake Superior Region: A Collection of Reference Articles* (Calumet: privately published, 1961), contains an extensive collection of articles culled from many sources relating to the Indian mining operations around Lake Superior. There is some mention of early reports of the presence of copper by white men and of the first attempts by European settlers to exploit the red treasure. See also Charles Whittlesey, "Ancient Mining on the Shores of Lake Superior," in *Smithsonian Contributions to Knowledge*, 13 (Washington, D.C., 1865).

16. George M. Wrong, ed., *Sagard's Long Journey to the Country of the Hurons*, Hugh Hornby Langton, trans. (Toronto: The Champlain Society, 1939), pp. 242–43.

17. Reuben Gold Thwaites, ed., *Travels and Explorations of the Jesuit Missionaries in New France, 1610–1791*, 73 vols. (Cleveland: The Burrows Brothers Co., 1896–1901), 50:265–67 (hereafter, *Jesuit Relations*).

18. Ibid., 54:153–65.

19. Ibid., 55:99.

20. For an overview of Talon's efforts to develop the trade of New France, see J. Bartlet Brebner, *Canada, A Modern History* (Ann Arbor: University of Michigan Press, 1960), pp. 46, 48–52.

21. Ernest Gagnon, *Louis Jolliet* (Quebec: Grande-Alle, 1902), pp. 15–18.

22. Francis Parkman, *La Salle and the Discovery of the Great West* (New York: Charles Scribner's Sons, 1915), pp. 48–56.

23. *N.Y. Records*, 9:344.

24. Francis Parkman, *A Half Century of Conflict*, 2 vols. (New York: Charles Scribner's Sons, 1915), 1:348–53; *Jesuit Relations*, 64:338.

25. *Jesuit Relations*, 69:39

26. *Calendar of State Papers, Colonial, American and West Indies, 1669–1674*, p. 232.

27. Lucius I. Barber, *A Record and Documentary History of Simsbury: 1643–1888* (Simsbury: The Commission for the Tercentenary, 1931), p. 182. Of the several published accounts that describe the Simsbury copper venture, Barber's is the most comprehensive and accurate, drawing heavily on the Simsbury Town Records for detailed accounts of events. See also John E. Ellsworth, *Simsbury, Being a Brief Historical Sketch of Ancient and Modern Simsbury, 1642–1935* (Simsbury: Simsbury D.A.R., 1935), pp. 109–22; J. M. French, "The Simsbury Copper Mines," *The New England Magazine*, Old Series, 5 (1887):427–37.

28. Barber, *Simsbury*, pp. 183–85.

29. Ibid., pp. 185–87.

30. Ibid.

31. Franklin Bowditch Dexter, *Sketch of the History of Yale University* (New York: Henry Holt and Co., 1887), p. 9.

32. *Conn. Records*, 5:104–05, 285.

33. Barber, *Simsbury*, pp. 191–97.

34. Thomas F. Gordon, *A Gazetteer of the State of New Jersey* (Trenton: D. Fenton, 1834), p. 11, sets the date of discovery as 1719. The earlier date and the legend of how the mine was discovered were reappraised by William Nelson, "Josiah Hornblower and the First Steam Engine in America," *Proceed-*

ings of the *New Jersey Historical Society*, 7 (1883):182 ff., and further adjusted downward to 1712, 1713, by Herbert P. Woodward, *Copper Mines and Mining in New Jersey*, Bull. 57, Geologic Series, Department of Conservation and Development, State of New Jersey (Trenton, 1944), pp. 39–69.

35. *N.J. Archives*, 5:7–8.

36. Ibid., 9; also Great Britain, Board of Trade, *Journal of the Commissioners for Trade and Plantations*, 14 vols. (London, 1920–38), 4:285–86; *Calendar of State Papers, Colonial, America and West Indies, 1720–1721*, p. 349.

37. William Cobbett, *Parliamentary History of England from the Earliest Period to the Year 1803*, 36 vols. (London: T. C. Hansard, 1806–20), 7:974, cclii. Cf. Charles M. Andrews, *The Colonial Period of American History: England's Commercial and Colonial Policy* (New Haven: Yale University Press, 1938), pp. 103–05, 105n; Leo Francis Stock, ed., *Proceedings and Debates of the British Parliament Respecting North America*, 5 vols. (Washington, D.C.: Carnegie Institution of Washington, 1924–41), 3:449n, 481. The enumeration of copper was renewed by the laws 11 George I c. 19, 8 George II c. 21, 15 George II c. 33, 20 George II c. 47, and 33 George II c. 16.

38. Gordon, *Gazetteer*, p. 14.

39. *N.J. Archives*, 5:267.

40. See Elizabeth Marting, "Arent Schuyler and His Copper Mine," *Proceedings of the New Jersey Historical Society*, 65 (1947):130–31, 131n.

41. There are no complete figures for the imports of copper, brass, and bronze for the early eighteenth century. The account given is my estimate based on a study of colonial coppersmithing (see Chapter 5).

42. *Calendar of State Papers, Colonial, America and West Indies, 1669–1674*, p. 232.

43. *Conn. Records*, 5:312–13.

44. Ibid., 6:84–87.

45. Charles Henry Stanley Davis, *History of Wallingford, Connecticut, from Its Settlement in 1670 to the Present* (Meriden: C. H. S. Davis, 1870), pp. 47–53.

46. *Conn. Records*, 6:371–72.

47. Barber, *Simsbury*, p. 197.

48. George F. Dow, "The Topsfield Copper Mines," *Proceedings of the Massachusetts Historical Society*, 65 (1936):570–80.

49. *Belcher Papers*, pt. 1, vol. 56; pt.2, vol. 57 (Boston: Massachusetts Historical Society, 1893–94), 56:33, 40, 80.

50. Richard H. Phelps, *Newgate of Connecticut: Its Origin and Early History* (Hartford: American Publishing Co., 1876), p. 19.

51. *Conn. Records*, 7:580–84.

52. Ibid., 14:93.

53. Ibid., pp. 122, 205–08. See also Richard H. Phelps's *Newgate History* and "The Newgate of Connecticut," *The Magazine of American History*, 15 (1886):321–24.

54. Benjamin Trumbull, *A Complete History of Connecticut*, 2 vols. (New Haven: Maltby, Goldsmith and Co., 1898), 1:25.

55. See Weiss and Weiss, *Old Copper Mines*, pp. 26–28; Woodward, *Mining in New Jersey*, pp. 70–72.

56. Weiss and Weiss, *Old Copper Mines*, pp. 30–39;

57. *N.J. Archives*, 19:264; 20:77, 400, 626; 24:452.

58. Weiss and Weiss, *Old Copper Mines*, pp. 63–79; Woodward, *Mining in New Jersey*, pp. 109–17.

59. Weiss and Weiss, *Old Copper Mines*, pp. 47–57.

60. When writing about the first American copper mining venture, J. Leander Bishop, *A History of American Manufactures from 1608 to 1860*, 3d ed., rev. and enl., 3 vols. (Philadelphia: E. Young, 1868), 1:509, referred to the existence of a British law that prohibited the smelting and refining of copper in the colonies. Other historians have followed Bishop's interpretation. The unhesitating actions of the colonists in erecting smelting works in New Jersey and Connecticut after the supposed passage of such a law tend to raise questions about its existence. In the present study, a careful review of English law back to the reign of William and Mary failed to find such a prohibition.

61. Charles P. Keith, "Sir William Keith," *The Pennsylvania Magazine of History and Biography*, 12 (1888):14, 16, 19.

62. Leonard W. Labaree, ed., *The Papers of Benjamin Franklin*, 21 vols. to date (New Haven: Yale University Press, 1959–), 3:465.

63. Nelson, "Josiah Hornblower," pp. 186–95.

64. Marting, "Arent Schuyler," p. 134; newspaper records of the fires are in *N.J. Archives*, 24:19 and 26:220, and the fires are described in a letter by Josiah Hornblower discovered by Richard P. McCormick, "The First Steam Engine in America," *Journal, Rutgers University Library*, 11 (1947):16–20.

65. *Penn. Archives*, 2:311-12.

66. J. Thomas Scharf, *History of Maryland*, 3 vols. (Baltimore: J. B. Piet, 1879), 1:437.

67. Robert Alonzo Brock, ed., *The Official Letters of Alexander Spotswood*, 2 vols. (Richmond: Virginia Historical Society, 1882–85), 1:162; Louis B. Wright and Marion Tinling, eds., *The Secret Diary of William Byrd of Westover: 1709–1712* (Richmond: The Dietz Press, 1941), p. 534.

68. William Byrd, *A Journey to the Land of Eden and Other Papers*, Mark Van Doren, ed. (New York: Macy-Masius, 1928), pp. 268–69, 272, 287, 303.

69. Ibid., p. 274. One of the first writers to use Byrd's writings to indicate the significance of copper mining in Virginia was Bishop, *American Manufactures*. Byrd's remarks are noted in 1:599.

70. Louis Morton, *Robert Carter of Nomini Hall* (Williamsburg: Colonial Williamsburg Inc., 1941), pp. 18–19; Fairfax Harrison, "The Will of Charles Carter of Cleve," *The Virginia Magazine of History and Biography*, 31 (1923):48–49 ff.; "Carter Papers," ibid., 6 (1898):18.

71. Alexander Henry, *Travels and Adventures in Canada and the Indian Territories between the Years 1760 and 1776* (Boston: Little, Brown, 1901), pp. 186–87, 195–97.

72. *N.Y. Records*, 8:140–42.

73. Sir William Johnson, *The Papers of Sir William Johnson*, 13 vols. (Albany: The University of the State of New York, 1921–62), 7:882–83.

74. Henry, *Travels*, pp. 212–13, 220–29.

75. Daniel J. Boorstin, *The Americans: The Democratic Experience* (New York: Random House, 1973), pp. 3–88.

76. Marting, "Arent Schuyler," pp. 135–40; Nelson, "Josiah Hornblower," pp. 225–26.

Chapter 4

1. Robert Alonzo Brock, "Early Iron Manufacture in Virginia: 1619–1776," in *Proceedings of the United States National Museum, 1885*, 8 (Washington, 1886):77.

2. National Park Service, U.S. Department of the Interior, *Saugus Iron Works*, pamphlet (Washington, 1975).

3. *Mass. Records*, 2:71.

4. Franklin Bowditch Dexter, ed., *New Haven Town Records, 1649–1684*, 2 vols. (New Haven, 1917), 1:235, 260, 349; Black, *John Winthrop*, pp. 173–76.

5. Charles Jeremy Hoadley, ed., *New Haven Colonial Records*, 2 vols. (Hartford: Case, Lockwood, 1857–58), 2:149, 173.

6. Herbert C. Keith and Charles Rufous Harte, "The Early Iron Industry of Connecticut," *Fifty-first Annual Report of the Connecticut Society of Civil Engineers* (1935), p. 8.

7. Dexter, ed., *New Haven Town Records*, 2:133–34, 138.

8. *Mass. Records*, 4, pt. 1:311–12.

9. Henry S. Nourse, ed., *The Early Records of Lancaster, Massachusetts: 1643–1725* (Lancaster: W. J. Coulter, 1884), pp. 48–49; Hartley, *Saugus*, pp. 276–79.

10. Hartley, *Saugus*, pp. 187–90.

11. Ibid., pp. 208–10.

12. Arthur Cecil Bining, *British Regulation of the Colonial Iron Industry* (Philadelphia: University of Pennsylvania Press, 1933), p. 13. There is some indication that Joshua Foote of Providence may have attempted to build a forge in that city just before his death in 1655 and that Daniel Jencks, Joseph's brother, was to have been apprenticed there. See Howard M. Chapin, "Proposed Iron Works at Providence in 1655," *Rhode Island Historical Society Collections*, 28 (1935):17–70. Foote was one of the three commissioners and attorneys who represented the Hammersmith ironworks in the lawsuits that followed its bankruptcy.

13. Alanson Borden, *Our County and Its People: A Description and Biographical Record of Bristol County, Massachusetts* (Boston: The Boston History Co., 1899), pp. 261–64

14. Charles E. Boyer, *Early Forges and Furnaces in New Jersey* (Philadelphia: University of Pennsylvania Press, 1931), pp. 196–99.

15. Hartley, *Saugus*, p. 275.

16. Joseph B. Felt, *Annals of Salem*, 2 vols. (Salem: W. and S. B. Ives, 1845), 1:282.

17. Bining, *British Regulation*, pp. 13–14.

18. Hartley, *Saugus*, p. 304.

19. *Calendar of State Papers, America and West Indies, 1675–1679*, p. 465.

20. Hartley, *Saugus*, pp. 202–07.

21. Ibid., pp. 14–20; *Mass. Records*, 4, pt. 1:217–20, 227, 228, 241–44, 251, 252.

22. Wilfred Smith, *An Historical Introduction to the Economic Geography of Great Britain* (New York: Frederick A. Praeger, 1968), pp. 106–08.

23. Thomas S. Ashton, *Iron and Steel in the Industrial Revolution* (Manchester: Manchester University Press, 1951), p. 104.

24. 3 George I c. 1; *Journals of the House of Commons* (1717), pp. 474, 478, 480–82, 486; *Journals of the House of Lords* (1717), pp. 421, 422. Details of the diplomatic correspondence are given in Cobbett, *Parliamentary History*, 7:395–421. See also Bining, *British Regulation*, pp. 36–38.

25. American papers reported the Russian depredations via dispatches received from London; see *Philadelphia American Weekly Mercury* for August 24 and October 5, 1721.

26. Michael Warren Robbins, "The Principio Company: Iron-Making in Colonial Maryland, 1720–1821" (Ph.D dissertation, The George Washington University, 1972), p. 14.

27. Henry Whitley, "The Principio Company," *The Pennsylvania Magazine of History and Biography*, 11 (1887):63.

28. Ibid., pp. 63–68; George Johnston, *History of Cecil County, Maryland* (Elkton: The Author, 1881), pp. 233–34; Robbins, "Principio Company," p. 26.

29. Benjamin Franklin, *Autobiography* (New York: Houghton Mifflin, 1886), p. 55; Kathleen Bruce, *Virginia Iron Manufacture in the Slave Era* (New York: The Century Co., 1930), pp. 15–16; Robbins, "Principio Company," pp. 35–37.

30. Whitley, "Principio," p. 192.

31. Keach Johnson, "The Baltimore Company Seeks English Markets: A Study of the Anglo-American Iron Trade, 1731–1755," *William and Mary Quarterly*, 3d series, 16 (1959):37–47.

32. See Joseph E. Walker, "Negro Labor in the Charcoal Iron Industry of Southeastern Pennsylvania," *Pennsylvania Magazine of History and Biography*, 93 (1969):466–86.

33. Whitley, "Principio," p. 64; Robbins, "Principio Company," pp. 192–95.

34. Whitley, "Principio," pp. 288–293. In 1785, claims brought against the confiscation of the Principio property were rejected by the governor and council because "Claimant and debtor both British subjects" (Roger Thomas, ed., *Calendar of Maryland State Papers: No. 3, The Brown Books* [Annapolis: The Hall of Records Commission, 1948], p. 124).

35. Brock, ed., *Alexander Spotswood*, 1:20–21, 41, 88–89.

36. Lester J. Cappon, *Iron Works at Tuball: Terms and Conditions for the Lease as Stated by Alexander Spotswood* (Charlottesville: University of Virginia Press, 1945), pp. 3–16.

37. The first German settlers consisted of the families arriving in 1714. Over the next ten years, others followed. See Elizabeth C. Denny Vann and Margaret C. Denny Dixon, *Virginia's First German Colony* (Richmond, 1961), pp. 9–34.

38. William Byrd, *A Progress to the Mines*, in *The Prose Works of William Byrd of Westover*, Louis B. Wright, ed. (Cambridge: Harvard University Press, 1966), pp. 356–58.

39. Ibid., pp. 357–58.

40. Bining, *British Regulation*, p. 35.

41. Ibid. The shifting opinions over the place of the British colonies in the overall economic life of Great Britain, which marked the opening of the eighteenth century, included considerable attention to the role of iron. Bining treats the early attempts to regulate the colonial iron industry in his Chapter 2, "The Opening Battle," pp. 32–48.

42. Bining, *Pennsylvania Iron*, pp. 49–51. The normal spelling for Darby's works was "Coalbrookdale."

43. Ibid., pp. 117–19.

44. Robbins, "Principio Company," pp. 23–88.

45. Bining, *Pennsylvania Iron*, pp. 117–18; Alfred Gemmell, "The Charcoal Iron Industry in the Perkiomen Valley," *Bulletin of the Historical Society of Montgomery County* [Penn.], 6 (1948):256.

46. New York and Philadelphia papers were the primary vehicles used in attempting to recruit skilled iron craftsmen. A typical example is the advertisement seeking experienced men to operate a new steel furnace in New York in 1775 cited in Rita Gottesman, ed., *The Arts and Crafts in New York, 1726–1776* (New York: New York Historical Society, 1938), p. 215.

47. Paul Paskoff, *Colonial Merchant-Manufacturers and Iron: A Study in Capital Transformation, 1725–1775* (Ph.D. dissertation, The Johns Hopkins University, 1976), pp. 14–17.

48. Keach Johnson, "The Genesis of the Baltimore Ironworks," *Journal of Southern History*, 19 (1953):176.

49. Paskoff, *Merchant-Manufacturers*, p. 72.

50. Ibid., p. 15, n. 27.

51. Byrd, *Prose Works*, p. 353.

52. Samuel Gustaf Hermelin, *Report about the Mines of the United States of America, 1783*, Amandus Johnson, trans. (Philadelphia: The John Morton Memorial Museum, 1931), p. 72.

53. Bining, *Pennsylvania Iron*, p. 75.

54. Ibid., pp. 29–35.

55. Starting with Hammersmith, ironworks bartered iron for a broad range of supplies that were dispersed to workers on a credit system. Typical examples of the practice are found in the ledgers for Colebrookdale Furnace, Coventry Iron Works, Pine Forge, and other Pennsylvania ironworks in the Potts Manuscript Collection, Historical Society of Pennsylvania, Philadelphia.

56. Byrd, *Prose Works*, pp. 352, 360.

57. Bruce, *Virginia Iron*, p. 17.

58. Letter of David Ross to John Hook, October 1776, David Ross Papers, Manuscript Collection, Virginia State Library, Richmond.

59. Robbins, "Principio Company," pp. 16–19, 223–26.

60. Paskoff, *Merchant-Manufacturers*, pp. 198–202, 217–20.

61. Bining, *Pennsylvania Iron*, Appendix, contains a comprehensive list of iron-producing facilities built in Pennsylvania before the Revolution.

62. The cast-iron, ten-plate cooking stove, an item that became a staple of the charcoal iron industry for nearly one hundred years, did not appear on the list of manufactured goods until 1767, when Thomas Maybury cast the first one at Hertford Furnace in Berks County, Pennsylvania. See H. Winslow Fegley, "Old Charcoal Furnaces Situated in Eastern Sections of Berks County," *Transactions of the Historical Society of Berks County*, 2 (1905):25–36.

Chapter 5

1. "Lord Cornbury to Mr. Secretary Hodges," July 15, 1705, *N.Y. Records*, 4:1151.

2. Carl Bridenbaugh, *The Colonial Craftsman* (New York: New York University Press, 1950), pp. 3–7. Bridenbaugh presents a broad survey of all of the crafts practiced during the colonial period, particularly during the eighteenth century. His treatment of metal manufacture, in consequence, is somewhat limited.

3. Bruce, *Economic History*, 2:146–47, 168–77. Pewter is an alloy of lead and tin. Because of its low melting point, it could be used on the table but not at the fire. Alchemy was primarily a copper-zinc alloy imitating gold in color, "alchemy gold," in reference to the goal of transmutation by alchemists.

4. Ibid., pp. 172–73. Silver artifacts often represented a form of savings in the absence of banks.

5. John Demos, *A Little Commonwealth: Family Life in Plymouth Colony* (New York: Oxford University Press, 1970), pp. 37–45.

6. George Francis Dow, ed., *The Probate Records of Essex County, Massachusetts*, 3. vols. (Salem: The Essex Institute, 1916–20), 1:456–57.

7. See estate lists of Jane Kenning of Ipswich and Rebecca Bacon of Salem, ibid., 1:164–65, 229–30.

8. Ibid., 1:330–31; 2:139–40, 319–20.

9. Bruce, *Economic History,* 2:147. The designation ten-penny signifies the number of nails, ten, that could be purchased for one penny. At times, nails actually circulated as currency in the specie-poor colonies, with the values indicated by the designation.

10. Arber and Bradley, eds., *John Smith,* 1:12–13.

11. Henry Wyckoff Belknap, *Trades and Tradesmen of Essex County, Massachusetts* (Salem: The Essex Institute, 1929), pp. 14, 31; Frederick Fairchild Sherman, *Early Connecticut Artists and Craftsmen* (New York: By the author, 1925), p. 29.

12. Bining, *British Regulation,* p. 13; Borden, *Bristol County, Massachusetts,* pp. 261–64.

13. Albert H. Sonn, *Early American Wrought Iron,* 3 vols. (New York: Charles Scribner's Sons, 1928), 1:12.

14. Bridenbaugh, *Colonial Craftsman,* p. 85.

15. George Francis Dow, *The Arts and Crafts in New England, 1705–1775* (Topsfield: The Wayside Press, 1927), reviewing the advertisements for various crafts in the Boston newspapers for the period, cites only eight references to blacksmithing, including such specialized variations as locksmith, gunsmith, and whitesmith. Over twenty silver- and goldsmiths advertised in the Boston papers during the same period.

16. Alston Deas, *The Early Ironwork of Charleston* (Columbia: Bostwick & Thornley, 1941), pp. 28-29.

17. Alex W. Bealer, *The Art of Blacksmithing* (New York: Funk & Wagnalls, 1969), is a comprehensive treatment of the tools and techniques of the blacksmith, with numerous line drawings to illustrate each facet of the art.

18. Bining, *Pennsylvania Iron,* p. 79, estimated that a forge operating at full capacity could convert upward of two tons of iron per week from bog ore.

19. The casting of anvils is recorded in the *Public Foundry Journal,* Richmond, for the period February 1, 1779, to June 19, 1781, fols. 49, 70, 88, in Division of Manuscripts, Virginia State Library, Richmond.

20. George W. Neible, "Account of Servants Bound and Assized before James Hamilton, Mayor of Philadelphia," *Pennsylvania Magazine of History and Biography,* 31 (1907):87, 95, 196, 198, 199.

21. In one three-week period, June 7–28, 1722, the *American Weekly Mercury* of Philadelphia carried advertisements seeking the return of indentured servants who had escaped from Colonel Spotswood's Virginia ironworks, from Principio, and from William Hunt, a Bucks County blacksmith.

22. Henry J. Kauffman, *Early American Ironware* (Rutland: Charles E. Tuttle, 1966), presents a survey of the scope of ironwork in the colonial and early national periods. A description of the products of the blacksmith is given, pp. 51–80.

23. James Iredell Papers, "Records of Port Roanoke, 1771–1776," in the Southern Historical Collection, University of North Carolina, Chapel Hill;

also, Ira Wilson Barber, Jr., *The Ocean-Borne Commerce of Port Roanoke, 1771–1776* (Master's thesis, University of North Carolina, Chapel Hill, 1931), p. 31, appendixes 1, 2. I am indebted to George Fore of Sanford, N.C., for calling these records to my attention.

24. Bridenbaugh, *Colonial Craftsman*, pp. 15–18.

25. Ibid., pp. 8–9, 20–22, discusses the problems of supply in the largely rural South.

26. Deas, *Early Ironwork*, p. 29.

27. Gottesman, ed., *Arts and Crafts in New York*, pp. 198–203.

28. Ibid., pp. 203–04. See also Kauffman, *American Ironware*, pp. 81–84. A smoke jack was a windmill-like device utilizing the heat rising from the fire to turn a roasting spit.

29. Joseph A. Goldenberg, *Shipbuilding in Colonial America*, published for The Mariners Museum, Newport News, Virginia (Charlottesville: University Press of Virginia, 1976). The extensive appendixes provide a comprehensive survey of the ports of origin and sizes of ships built in the colonies throughout much of the eighteenth century.

30. Sir Westcott Abell, *The Shipwright's Trade* (Cambridge, England: The University Press, 1948), pp. 86–90.

31. Goldenberg, *Shipbuilding*, pp. 16–17, 95.

32. One of the first such forges was the Curtis Anchor Forge, erected in 1710 near Abington, Massachusetts. Forges capable of making large anchors had to await the development of colonial furnaces (after 1730) to cast the weights of metal necessary. John Barnard Pearse, *Concise History of the Iron Trade of the American Colonies up to the Revolution and of Pennsylvania until the Present Time* (Philadelphia: Allen, Lane and Scott, 1876), p. 28.

33. William Avery Baker, *A Maritime History of Bath, Maine and the Kennebec River Region*, 2 vols. (Bath: Marine Research Society of Bath, 1973), 1:93.

34. Gertrude Selwyn Kimball, *Providence in Colonial Times* (Boston: Houghton Mifflin, 1912), pp. 293–94.

35. Shirley Ann Martin, *Craftsmen of Bucks County, Pennsylvania, 1750–1800* (Master's thesis, University of Delaware 1956), pp. 108–50; also Kauffman, *American Ironware*, p. 62.

36. "Letter of Morgan Evan," *Pennsylvania Magazine of History and Biography*, 42 (1918):176–77.

37. *Blacksmith Daybook, 1742–1767*, Colebrookdale Furnace Collection (Potts MSS), fols. 74, 128, Historical Society of Pennsylvania, Philadelphia.

38. George Shumway and Howard C. Frey, *The Conestoga Wagon, 1750–1850*, 3d ed. (York: Geo. Shumway, 1968), is a well-illustrated history of the development and the influence of the Conestoga wagon on the expansion of the country since its invention. Excellent line drawings and photographs illustrate the many parts fabricated by the wheelwright and blacksmith, pp. 164–213.

39. James H. Craig, *The Arts and Crafts in North Carolina: 1699–1840* (Winston-Salem: Old Salem, Inc., 1965), pp. 128–40.

40. *Md. Archives*, 11:65.

41. Henry J. Kauffman, *The Pennsylvania-Kentucky Rifle* (Harrisburg: The Stackpole Company, 1960), colonial historical development, pp. 8–17; techniques of the gunsmith, pp. 141–68.

42. Bruce, *Economic History*, 2:170–74.

43. Dow, *Arts and Crafts in New England*, p. 65; John Marshall Phillips, *American Silver* (New York: Chanticleer Press, 1949), pp. 11–13.

44. Graham Hood, *American Silver: A History of Style, 1650–1900* (New York: Praeger Publishers, 1971), contains a good, recent bibliography that lists general books on early American silver and publications describing the silver manufacture of individual states, pp. 247–50. Although the emphasis is on changing tastes and styles in silverware, Hood also offers an account of the growth of silversmithing as a reflection of historical attitudes and events.

45. C. Louise Avery, *Early American Silver* (New York: The Century Co., 1930), pp. 11–16, 20–23.

46. *Mass. Records*, 3:261–62.

47. Samuel Eliot Morison, *Builders of the Bay Colony* (Boston: Houghton Mifflin, 1930), pp. 279–80.

48. Herman Frederick Clarke, "The Craft of Silversmith in Early New England," *New England Quarterly*, 12 (1939):68–79.

49. Herman Frederick Clarke, "John Hull: Mintmaster," *New England Quarterly*, 10 (1937):668–84, gives a concise history of the operation of the mint.

50. Avery, *Early American Silver*, pp. 124–29.

51. J. Hall Pleasants and Howard Sill, *Maryland Silversmiths: 1715–1830* (Baltimore: Lord Baltimore Press, 1930), p. 207.

52. "Broadnax Family," *William and Mary Quarterly*, 1st series, 21 (1912–13) 267.

53. George Barton Cutten, *Silversmiths of Virginia* (Richmond: The Dietz Press, 1952), pp. 2 f., 37 f., 81 f., 186 f.

54. George Barton Cutten, *The Silversmiths of North Carolina* (Raleigh: State Department of Archives and History, 1948), pp. 3–4.

55. E. Milby Burton, *South Carolina Silversmiths, 1698–1860* (Charleston: The Charleston Museum, 1942), p. 210.

56. George Munson Curtis, *Early Silver of Connecticut and Its Makers* (Meriden: International Silver Co., 1913), pp. 14–20. Henry J. Kauffman, *The Colonial Silversmith* (Camden: Thomas Nelson & Sons, 1969), is a well-illustrated description of the products and technique of the smith.

57. Phillips, *American Silver*, pp. 14–16.

58. Martha Gandy Fales, *Early American Silver* (New York: E. P. Dutton, 1973), pp. 4–31.

59. Avery, *Early American Silver*, pp. 183–86; Herman Frederick Clarke, *John Coney, Silversmith: 1655–1722* (Cambridge: The Riverside Press, 1932), pp. 3–17.

60. Harrold E. Gillingham, "The Library: Indian Trade Silver Ornaments Made by Joseph Richardson, Jr.," *Pennsylvania Magazine of History and Biography*, 67 (1943):83–88.

61. Gottesman, ed., *Arts and Crafts in New York*, pp. 35, 37–39.

62. Phillips, *American Silver*, pp. 12–13.

63. Elizabeth B. Potwine, "John Potwine: Silversmith of Connecticut and Massachusetts," *Antiques*, 28 (1935):106–09.

64. Peter Bohan and Philip Hammerslough, *Early Connecticut Silver: 1700–1840* (Middletown: Wesleyan University Press, 1970), p. 6.

65. Dow, *Arts and Crafts in New England*, p. 55.

66. Brass is defined as "any of the many copper-base alloys in which zinc is the principal alloying element"; bronze, "any of the many copper-base alloys in which tin is the principal alloying element" (in modern practice, any alloy in which the principal alloying element is any element other than zinc) (John Goulding Henderson and Jack M. Bates, *Metallurgical Dictionary* [New York: Reinhold, 1953], pp. 41, 47).

67. Henry J. Kauffman, *American Copper and Brass* (Camden: Thomas Nelson & Sons, 1968), is the most complete study of the art of coppersmithing in early America. A detailed discussion of the products and techniques is given, together with a documented list of all coppersmiths and brass founders to the Civil War period.

68. Dow, *Arts and Crafts in New England*, p. 226.

69. The colonial coppersmith shares this relative anonymity with the colonial blacksmith. With the exception of a few outstanding individuals, such as Paul Revere and David Rittenhouse, whose fame also rests on achievements other than their craftsmanship, very little attention has been paid to the colonial metalworkers in general, either regarding their overall contributions to history or their individual careers. Kauffman devotes one chapter to William Bailey of Yorktown (*American Copper and Brass*, pp. 251–60). Another interesting study of a colonial coppersmith using archaeological evidence is Ivor Noel Hume, *James Geddy and Sons: Colonial Craftsmen* (Williamsburg: Colonial Williamsburg Archaeological Series, No. 5, 1970). The subject remains a fertile area for future research.

70. The many references to alcoholic beverages in the colonial records leads to the inevitable conclusion that they were hard-drinking times. Stanley Baron, *Brewed in America* (Boston: Little, Brown, 1962), is the only serious study on the subject. His comprehensive survey of brewing during the colonial period (pp. 18–94) gives some idea of how extensive the practice was. As Baron records (pp. 95–100), the many recipes of the time all called for the boiling of the mash in a "Copper." The importance of rum, socially, commercially, and politically, to the colonies is the subject of Charles William Taussig's *Rum, Romance and Rebellion* (New York; Minton, Balch & Co., 1928). Here, stills would have had extensive use although Taussig does not discuss the technology.

71. Kauffman, *American Copper and Brass*, pp. 84–85.

72. Henry Hamilton, *The English Brass and Copper Industries to 1800*, 2d ed. (London: Frank Cass and Co., Ltd., 1967), pp. 342–44.

73. Johann David Schoepf, *Travels in the Confederation*, Alfred J. Morrison, trans., 2 vols. (Original translation from 1788 edition, 1911; reprinted, New York: Burt Franklin, 1968), 2:25–27.

74. Maxwell Whiteman, *Copper for America* (New Brunswick: Rutgers University Press, 1971), pp. 3, 47–48. The greatest name in early American copper manufacture was Paul Revere. He founded his rolling mill in Canton, Massachusetts, to make sheet for ship sheathing in 1801. See Esther Forbes, *Paul Revere and the World He Lived In* (Boston: Houghton Mifflin Co., 1944), pp. 407–12. John H. Morrison, *History of New York Shipyards* (1909; reissued, Port Washington; Kennikat Press, 1970), notes that the *Empress of China*, built in Baltimore in 1784, was one of the first American ships with copper sheathing. Before 1801, copper sheathing used by American yards was bought from England, including the sheathing on the original six frigates for the American navy (pp. 18–19).

75. John Raymond Harris, *The Copper King* (Toronto: University of Toronto Press, 1964), p. 12.

76. Hamilton, *English Brass and Copper*, p. 342.

77. William G. Lathrop, *The Brass Industry in Connecticut* (Shelton: By the author, 1909), p. 38; Kenneth T. Howell, *History of Abel Porter & Company* (Waterbury: Scovill Manufacturing Co., 1952), pp. 13–14.

78. Neible, "Account of Servants Bound," p. 94.

79. J. Thomas Scharf and Thompson Westcott, *History of Philadelphia*, 3 vols. (Philadelphia: L. H. Evarts, 1884), 1:244–45, 245n.

80. Richard D. Moore, "The Higley Coppers, 1737–1739," *Connecticut Historical Society Bulletin*, 20 (1945):69–73.

81. See Eugene S. Ferguson, ed., *Early Engineering Reminiscences (1815–40) of George Escol Sellers* (Washington, D.C.: United States National Museum Bulletin 238, 1965), pp. 66–67 ff.

82. Silvio A. Bedini, *Early American Scientific Instruments and Their Makers* (Washington, D.C.: United States National Museum Bulletin 231, 1964), is a detailed study of scientific instruments through the colonial period which lists about two hundred men who were engaged in their manufacture.

83. Brooke Hindle, *David Rittenhouse* (Princeton: Princeton University Press, 1964), pp. 72–75, 87; Howard C. Rise, Jr., *The Rittenhouse Orrery* (Princeton: Princeton University Press, 1954), pp. 28–36.

84. Ledlie I. Laughlin, *Pewter in America*, 2 vols. (Boston: Houghton Mifflin, 1940), contains the most comprehensive treatment of colonial pewter, including a description of the products of the craft. Volume 2 has an extensive bibliography on all aspects of the craft in America, pp. 163–92.

85. H. J. L. J. Masse, *Chats on Old Pewter* (London: T. Fisher Unwin, 1911), pp. 107–31, discusses at length the various grades of pewter and their compositions and the several fabrication techniques by which pewter objects were and are made. Henry J. Kauffman, *The American Pewterer: His Techniques and His Products* (Camden: Thomas Nelson & Sons, 1970), contains excellent line drawings illustrating the production of pewterware.

86. George Francis Dow, "Notes on the Use of Pewter in Massachusetts during the Seventeenth Century," *Old-Time New England*, 14 (1923):29–33, claims one Richard Graves of Salem to be the first to practice the trade of pewterer in the colonies, at times, between the years 1642 and 1655.

87. Laughlin, *Pewter in America*, 1:5–6.

88. R. W. Symonds, "The English Export Trade in Furniture to Colonial America," *Antiques*, 28 (1935):156.

89. John Barrett Kerfoot, *American Pewter* (New York: Crown, 1924), gives a detailed history of all known American pewterers and their locations, including tables of pewterers by city, number, and period of occupation, confirmed or suspected, pp. 34–71.

90. Madelaine R. Brown, "Rhode Island Pewterers," *Rhode Island Historical Society Collections*, 31 (1938):1–8.

91. R. Malcolm Keir, "The Tin-Peddler," *Journal of Political Economy*, 21 (1913):255–58.

92. Shirley Spaulding DeVoe, *The Tinsmiths of Connecticut* (Middletown: Wesleyan University Press, 1918), pp. 3–6, 12–15.

93. Jackson Turner Main, *The Social Structure of Revolutionary America* (Princeton: Princeton University Press, 1965), pp. 67, 75–83, 112–13, 132–35, 199, 219. Main considers the general class of "artisan," which includes metals craftsmen, but there is enough differentiation of trades to show the slightly elevated position of metalworkers within the class.

94. As late as 1810, overland freight costs were as high as 20 to 60 cents a ton per mile, a serious barrier to the transportation of iron goods for any distance (Victor S. Clark, *History of Manufactures in the United States*, 3 vols. [New York; McGraw-Hill, 1929], 1:337–38).

1. Cited by Archibald Kennedy in *Observations on the Importance of the Northern Colonies* (New York: James Parker, 1750), p. 11.

2. Ibid., p. 12.

3. *Mass. Records*, 4, pt. 1:311–12.

4. The economic philosophy of mercantilism held that colonies were to serve primarily as sources of raw materials to and markets for manufactured goods from the mother country. Victor S. Clark, who fully subscribed to the mercantile theory as the motive for the founding of the American colonies by England, gave a detailed description of how it was to have functioned in the colonies, *History of Manufactures*, 1:9–12.

5. See John Trenchard, *Cato's Letters*, 4 vols. (London: Printed for W. Wilkins, T. Woodward, J. Walthoe, J. Peele, 1723–24), 4:286, 290.

6. *Md. Archives*, 33:467–69.

7. *Conn. Records*, 6:312–13.

8. Ibid., 7:174–75.

9. Ibid., 8:338–39; 9:58.

10. *R. I. Records*, 4:295.

11. Ibid., 5:17; 6:573–74.

12. *Penn. Records*, 4:52, 247–48, 266–73.

13. *N.Y. Records*, 6:116–17.

14. Bining, *British Regulation*. Bining surveys the full range of colonial records and government documents to establish a closely reasoned account of the interrelationships between the growing industry and British and American policies in the colonial period.

15. Ibid., pp. 39–43.

16. Ibid., p. 43.

17. Ibid., pp. 51–52.

18. British iron suffered from sulfur and phosphorus contamination adversely affecting its performance in some applications. The quality of the Swedish iron was based on freedom from those impurities. American iron approached the Swedish in quality, although it was considered slightly inferior, probably by the prejudice of tradition, on both sides of the Atlantic. See Theodore Wertime, *The Coming of the Age of Steel* (Chicago: University of Chicago Press, 1962), pp. 22–26.

19. Harry Scrivenor, *History of the Iron Trade* (London: Longman, Brown, Green and Longmans, 1854), p. 72.

20. *Journals of the House of Commons* (1738), p. 157.

21. Ibid., p. 172; Bining, *British Regulation*, pp. 54–62; Ashton, *Iron and Steel*, pp. 116–18.

22. Morristown (originally New Hanover) and the surrounding area were settled in 1710 by Puritans attracted to the iron deposits there. By the Revolution, twenty or more forges and furnaces were operating in the county. Many of the works are described in Boyer, *Early Forges and Furnaces*, p. 33 and passim.

23. "Governor Lewis Morris to the Lords of Trade, December, 1741," in *The Papers of Governor Lewis Morris*, New Jersey Historical Society Collections, 12 vols. (Trenton: The Society, 1846–), 4:141–42.

24. Gottesman, ed., *Arts and Crafts in New York*, p. 212.

25. James M. Ransom, *Vanishing Ironworks of the Ramapos* (New Brunswick: Rutgers University Press, 1966), pp. 177–214, recounts the long, colorful history of the Sterling works, which operated almost continuously until 1923.

26. Bining, *British Regulation*, pp. 64–68; Ashton, *Iron and Steel*, pp. 118–20.

27. Anon., "The State of the Trade and Manufactory of Iron in Great Britain Considered" (1750), in the Manuscript Division, Library of Congress, Washington, D.C., pp. 8, 15.

28. Bining, *British Regulation*, p. 68.

29. Pickering, *Statutes*, 20:97–102.

30. Ashton, *Iron and Steel*, pp. 119–21. See also Scrivenor, *Iron Trade*, pp. 76–80, for a discussion of the debate over the removal of the London restriction.

31. A slitting mill's function was to cut plate iron into thin strips from which nails were cut by hand. A plating forge or mill hammered bar iron into flat plate.

32. Pickering, *Statutes*, 20:100.

33. Bining, *British Regulation*, p. 83.

34. Scrivenor, *Iron Trade*, Appendix, p. 343.

35. *Conn. Records*, 10:623.

36. *Historical Statistics of the United States: Colonial Times to 1957* (Washington, D.C.: U.S. Department of Commerce, 1960), pp. 762–63. In 1772, Virginia and Maryland shipped to England 1,873 long tons of pig iron out of a total of 3,725 tons for all the colonies. Pennsylvania contributed only 706 tons. Total exports of bar iron for 1772 came to about 900 tons. Iron sent down the Susquehannna River from Pennsylvania to Maryland ports for export was credited to Maryland in the reports.

37. "Roger Wolcott to Commissioners for Trade and Plantations, May 28, 1751," *Connecticut Historical Society Collections*, 31 vols. (Hartford: The Society, 1860–1932), 16:74–75.

38. *N.J. Archives*, 7:258–59.

39. *N.Y. Records*, 6:604–05.

40. See Bining, *British Regulation*, Appendix A, pp. 126–27.

41. Ibid., p. 73, for a summary of reports on the 1750 census.

42. Ibid., pp. 86–87.

43. Gottesman, ed., *Arts and Crafts in New York*, p. 212.

44. "Philip Livingston to Roger Wolcott, February 12, 1745/46," *Connecticut Historical Society Collections*, 13:186.

45. Bining, *British Regulation*, pp. 89–91.

46. *N.J. Archives*, 10:31; 25:484.

47. Gottesman, ed., *Arts and Crafts in New York*, p. 215.

48. *N.J. Archives*, 26:62; 28:82, 177.

49. The events surrounding the imposition of the Stamp Act and the bitter reaction created in the colonies is fully treated by Edmund S. Morgan and Helen M. Morgan, *The Stamp Act Crisis* (Chapel Hill: University of North Carolina Press, 1953). The Morgans briefly note the enumeration of iron (p. 39) but make no mention of the larger issues of regulation. By the terms of British law, enumerated commodities produced in the colonies could be shipped only to England. Enumeration constituted a restriction on trade, not production.

50. William S. Taylor and John Henry Pringle, eds., *Correspondence of William Pitt, Earl of Chatham*, 4 vols. (London: John Murray, 1838–40), 2:373.

51. See Bining, *British Regulation*, p. 97, 97n.

52. John Dickinson, *Letters from a Farmer in Pennsylvania to the Inhabitants of the British Colonies* (Philadelphia: David Hall and William Sellers, 1768), p. 12.

53. City of Boston, *Report of the Record Commissioners Containing the Eighteenth Boston Town Records, 1770–1777* (Boston: Rockwell and Churchill, 1887), pp. 94–106.

54. Ibid., p. 104.

55. Thomas Jefferson, "A Summary View of the Rights of British America," *The Papers of Thomas Jefferson*, Julian P. Boyd, ed., 19 vols. to date (Princeton: Princeton University Press, 1950–), 1:125.

56. *Conn. Records*, 13:617.

57. *R.I. Records*, 7:242–43.

58. *N.C. Records*, 7:898, 899, 901, 903, 937, 940, 941, 947, 949; 8:166, 196.

59. Ibid., 10:216–19.

60. Bining, *Pennsylvania Iron*, p. 59.

61. Arthur D. Pierce, *Iron in the Pines* (New Brunswick: Rutgers University Press, 1957), pp. 20–29. Pierce gives a detailed history of the founding and development of the pine barrens iron industry from Read's efforts to its gradual collapse after the middle of the nineteenth century.

62. Boyer, *Early Forges and Furnaces*, pp. 154–58.

63. Pierce, *Iron in the Pines*, pp. 66, 123–26.

64. Hasenclever wrote a highly favorable account of his activities in 1773 to refute his critics after the collapse of the American Iron Company entitled *The Remarkable Case of Peter Hasenclever, merchant: Formerly one of the Prospectors of the Iron Works Pot-Ash Manufactory, & c. established and successfully carried on under his Direction, in the Provinces of New York, and New Jersey in North America, 'till November 1766* (London, 1773); copy in the Manuscript Division, Library of Congress.

65. Ibid., p. 3. In the period just before the Revolution, other Englishmen continued to find the forest wealth of America attractive. Richard Jackson of England wrote to Jonathan Trumbull of Connecticut in 1767 concerning a proposal to produce iron in the colony since the colonists lacked commoditites to sell to England in exchange for manufactured goods, "Iron Ore is wanting no where, great Command of water we (England) have in many places the only advantage you have over us is in your plenty of fuel" ("Richard Jackson to Jonathan Trumbull," October 19, 1767, *Connecticut Historical Society Collections*, 19:59).

66. Hansenclever, *Peter Hansenclever, merchant*, p. 5.

67. Ibid., pp. 6–8; Ransom, *Vanishing Ironworks*, pp. 17–19.

68. See the report of the "Franklin Commission" which investigated the soundness of Hasenclever's works: "He is also the first we know of, who has rendered the old cinder beds of the furnaces useful and profitable" (*N.J. Archives*, 28:252).

69. Ransom, *Vanishing Ironworks*, p. 70. Hasenclever's pamphlet defending his position drew an angry rebuttal in the New York press from his replacement, Robert Erskine. Erskine accused the German of gross fiscal mismanagement (*N.J. Archives*, 28:586–92).

70. *N.J. Archives*, 9:584, 584n.

71. Hasenclever, *Peter Hasenclever, merchant*, pp. 8–9.

72. Ibid., p. 26.

73. *N.J. Archives*. The entire report is printed in 7:247–53.

74. Ransom, *Vanishing Ironworks*, pp. 17–27.

75. Quoted in Gerhard Spieler, "Peter Hasenclever, Industrialist," *Proceedings of the New Jersey Historical Society*, 59 (1941):254.

76. *Historical Statistics*, pp. 763–85; *N.Y. Records*, 5:609.

77. Iredell, "Records of Port Roanoke."

78. Bruce, *Virginia Iron*, pp. 65–66, 454; Brock, "Early Iron Manufacture in Virginia," p. 80.

79. David Ramsay, *History of South Carolina*, 2 vols. (Newberry: W. J. Duffie, 1858), 2:Appendix, 307.

80. Bining, *British Regulation*, pp. 26–27, summarizes the data on charcoal blast furnaces from a number of sources dating from 1757 to the Revolution. The exact number of furnaces constructed before 1775 probably was larger than the eighty-two Bining claims, since a number were constructed in some colonies after the dates of the reports he used, making an exact compilation extremely difficult with the data available. Another problem is the uncertainty as to the number that ceased operation before 1775 or were in operation in any one year. Eighty is a reasonable estimate of the number of furnaces capable of operation in 1775.

81. Hermelin, *Mines of the United States*, pp. 59–60, 69, 72.

82. The figure of 175 forges is given by Bining, *British Regulation*, p. 29. The accuracy of the figure is subject to the same general qualifications noted for the number of furnaces (see note 80).

83. Ibid., pp. 26–27, 29–30, 122; Scrivenor, *Iron Trade*, pp. 86–88. A shift to coke, which had less tendency to crumble in the furnace than charcoal, permitting the English to build larger furnaces with increased efficiency and yield in the last quarter of the century, enabled them then to outproduce the American industry with fewer furnaces, but that transition had not been completely effected at the start of the Revolution.

Chapter 7

1. "Letter of Col. Henry Knox to Robert Treat Paine, June 24, 1776," in *Naval Documents*, 5:711.

2. For a comprehensive discussion of the evolution of American political ideology in this period, see Bernard Bailyn, *Ideological Origins of the American Revolution* (Cambridge: Harvard University Press, 1967). The theme is further elaborated in Part I of Gordon S. Wood, *The Creation of the American Republic, 1776–1787* (Chapel Hill: University of North Carolina Press, 1969), pp. 1–124.

3. John Adams, *The Works of John Adams*, Charles Francis Adams, ed., 10 vols. (Boston: Little, Brown, 1850–1856), 10:197.

4. At the same time, Parliament passed the "coercive" acts, they passed the Quebec Act extending the boundaries of Canada south to the Ohio River and giving nominal recognition to the practice of Roman Catholicism in the province. Although stemming from long-standing and unrelated issues, the colonists also considered the Quebec Act as "intolerable."

5. The key phrases in the Suffolk Resolves as endorsed by the Continental Congress on September 18, 1774, are reprinted in Henry Steele Commager and Richard B. Morris, eds., *The Spirit of 'Seventy-Six*, 2 vols. (New York: Bobbs-Merrill, 1958), 1:53–55; see also Edmund Cody Burnett, *The Continental Congress* (New York: The Macmillan Co., 1941), pp. 42–45.

6. A copy of the Dartmouth letter is printed in *N.Y. Records*, 8:509.

7. *R.I. Records*, 7:305.

8. "Letter of Captain Wallace to Vice-Admiral Samuel Graves, Dec. 14, 1774," in *Naval Documents*, 1:15. See also *Newport Mercury* of December 12, 1772, ibid., p. 14.

9. *R.I. Records*, 7:262.

10. Ibid., 7:306.

11. Otis G. Hammond, ed., *Letters and Papers of Major-General John Sullivan*, 3 vols. (Concord: New Hampshire Historical Society, 1930–39), 1:7–8; Jeremy Belknap, *The History of New Hampshire* (Dover: S. C. Stevens and Ela S. Wadleigh, 1831), p. 353.

12. Two of these survived the war and later were inscribed with the names "Hancock" and "Adams" and the motto "Sacred to Liberty" at the direction of Congress. See Bishop, *American Manufactures*, 1:487 ff.; "May 19, 1788," *Journals of the American Congress: From 1774 to 1788*, 4 vols. (Washington, D.C.: Way and Gideon, 1823), 4:816.

13. *The Journals of Each Provincial Congress of Massachusetts in 1774 and 1775 and of the Committee of Safety* (Boston: Dutton and Wentworth, 1838), p. 30.

14. Ibid., pp. 505–08.

15. Samuel Green Arnold, *History of Rhode Island, 1700–1790*, 2 vols. (New York: D. Appleton, 1860), 2:344.

16. "Letter of Vice-Admiral Graves to Philip Stephens, Secretary of the British Admiralty, April 11, 1775," *Naval Documents*, 1:178.

17. "Letter of Nathaniel Shaw, Jr. to Eliphalet Dyer, Dec. 14, 1774," ibid., 1:20.

18. "Letter of Vice-Admiral Samuel Graves to Captain George Vandeput, May 1, 1775," ibid., 1:255.

19. "Letter of Captain Vandeput to Vice-Admiral Graves, August 24, 1775," ibid., 1:1223–24.

20. Bruce Bliven, Jr., *Under the Guns: New York, 1775–1776* (New York: Harper and Row, 1972), pp. 34–39, 101–02, 142–44, 169–72, 292–94.

21. *R.I. Records*, 7:371; Arnold, *Rhode Island*, p. 365.

22. *R.I. Records*, 7:417, 464.

23. *Conn. Records*, 15:201, 224, 234, 242, 457, 484, 490.

24. Charles Jeremy Hoadley and Leonard Woods Labaree, eds., *Public Records of the State of Connecticut*, 9 vols. (Hartford, 1894–1953), 1:55–56.

25. *Penn. Records*, 10:412, 432, 561; *Penn. Archives*, 1st series, 4:721, 761, 762; 5:36, 62.

26. *Penn Records*, 11:7.

27. For a general description of the manufacture of cannon in the eighteenth century, see Alfred Rupert Hall, "Military Technology," *A History of Technology*, Charles Singer, Eric John Holmyard, Alfred Rupert Hall, and Trevor I. Williams, eds., 5 vols. (Oxford: Oxford University Press, 1954–58), 3:360–73.

28. Louis de Tousard, *American Artillerist's Companion*, 2 vols. (1809; reprint ed., New York: Greenwood Press, 1969), 1:xviii-xxi, 361–63.

29. Thomas M. Doerflinger, "Hibernia Furnace during the Revolution," *New Jersey History*, 90 (1972):106.

30. *Md. Archives*, 11:142, 167, 174, 180, 187, 235.

31. Ibid., 12:40, 149, 204, 337, 408, 440, 522.

32. Ibid., 11:134, 142, 167, 174, 180, 187; 12:11, 40, 73, 149.

33. Bruce, *Virginia Iron*, pp. 46–63, 76–79.

34. *N.C. Records*, 10:540, 869; Gottesman, ed., *Arts and Crafts in New York*, pp. 215–16.

35. David Duncan Wallace, *South Carolina: A Short History* (Chapel Hill: University of North Carolina Press, 1951), pp. 454–55.

36. Bining, *British Regulation*, p. 111.

37. *Journal of the Proceedings of Congress, May 10, 1775 to August 1, 1775* (London: J. Almon, 1776), p. 162.

38. *Journal of the Proceedings of Congress, September 5, 1775 to April 30, 1776* (London: J. Almon, 1778), p. 31.

39. *Journals of Congress*, 4:223–24.

40. *Journal of the Proceedings of Congress, September 5, 1775 to April 30, 1776*, pp. 103, 134, 135, 138.

41. Jared Sparks, ed., *Diplomatic Correspondence of the American Revolution*, 12 vols. (Boston: N. Hale and Gray & Bowen, 1829–30), 1:7.

42. Edward F. Curtis, *The Organization of the British Army in the American Revolution* (New Haven: Yale University Press, 1926), pp. 3–8.

43. John Burgoyne, *A State of the Expedition from Canada, as laid before the House of Commons, by Lieutenant-General Burgoyne, and verified by Evidence* (London: J. Almon, 1780), pp. 8–11.

44. Henry Dearborn, *The Revolutionary War Journals of Henry Dearborn, 1775-1783*, Lloyd A. Brown and Howard H. Peckham, eds. (Chicago: The Caxton Club, 1939), p. 127.

45. Francis S. Drake, *The Life and Correspondence of Major-General Henry Knox* (Boston: Samuel G. Drake, 1873), pp. 22–24, 129–30. The brass mortar was the largest piece of ordnance used in the war by either side. See *Naval Documents*, 2:1247.

46. Dearborn, *War Journals*, p. 221.

47. George Washington, *The Writings of George Washington*, John C. Fitzpatrick, ed., 39 vols. (Washington, D.C.: 1931–44), 4:59, 370–71, 404, 407, 438; 5:294 (hereafter, Washington, *Writings*).

48. *Journals of Congress*, 1:378, 451, 553, 568.

49. "Letter of Col. Henry Knox to Robert Treat Paine, June 24, 1776," *Naval Documents*, 5:710–12.

50. Ibid., 2:203, 431, 441; Washington, *Writings*, 6:474.

51. Washington, *Writings*, 7:18–22, 146–47.

52. Ibid., 8:110; *Journals of Congress*, 1:545.

53. Fusil: a light flintlock musket. See Sparks, *Diplomatic Correspondence*, 1:35, 61; Washington, *Writings*, 8:237; 9:181.

54. "Report of General Horatio Gates, February, 1778," in Records of the Continental and Confederate Congresses and the Constitutional Convention, Record Group 360, National Archives Microfilm Publication M247, Roll 47, 1, fol. 506, National Archives, Washington, D.C. (hereafter, Record Group 360).

55. "An account of shot and shells contracted for, for the use of the present campaign," August 15, 1780, in War Department Collection of Revolutionary War Records, Record Group 93, National Archives Microfilm Publication M859, Roll 65, item 20542, National Archives, Washington, D.C. (hereafter, Record Group 93).

56. Ransom, *Vanishing Ironworks*, pp. 137–39; *N.J. Archives*, 2d series, 5 vols. (Trenton: J. L. Murphy, 1901–17), 2:269.

57. *Penn. Archives*, 1st series, 6:170–71.

58. "Letter of John Jacob Faesch to Samuel Hodgdon, August 8, 1780," Record Group 93, Roll 72, item 21621.

59. Ransom, *Vanishing Ironworks*, pp. 38–50, 74–75, 79–80; Washington, *Writings*, 7:464–65.

60. *Journals of Congress*, 2:39.

61. Washington, *Writings*, 20:152, 448.

62. An alphabetical listing of all ships commissioned as privateers by the states or by Congress is given in Library of Congress, *Naval Records of the American Revolution, 1775–1788* (Washington: 1906).

63. Gardner W. Allen, *A Naval History of the American Revolution* (New York: Russell and Russell, 1962), pp. 596–97.

64. William M. Fowler, Jr., *Rebels under Sail: The American Navy during the Revolution* (New York: Charles Scribner's Sons, 1976), p. 281; John Richard Alden, *The American Revolution* (New York: Harper & Row, 1954), pp. 205–06.

65. See *Naval Documents*, 2:1247; 3:48.

66. *Journals of Congress*, 3:311–12.

67. "Letter from Ezek Hopkins to John Hancock, April 9, 1776," Record Group 360, Roll 78, 9, fol. 33.

68. "Letter of Gov. Trumbull [Conn.] to Continental Congress, April 27, 1776," ibid., Roll 66, fol. 165.

69. "Letter of Daniel Joy to the Cannon Committee of the Continental Congress, May 22, 1776," on the proving of cannon at Reading Furnace (*Naval Documents*, 5:207). See also Hindle, *David Rittenhouse*, pp. 125–26, 135; Marion Vernon Brewington, "American Naval Guns, 1775–1785," *The American Neptune*, 3 (1943):12.

70. "An Estimate of Cannon Ball and Ballast for the Continental Frigates Building at Poughkeepsie, March 8, 1776," *Naval Documents*, 4:233–34.

71. American furnaces lacked the capacity to cast cannon larger than eighteen-pounders (Brewington, "American Naval Guns," p. 13). For the luckless saga of the *Raleigh*, see the correspondence of John Langdon cited in *Naval Documents*, 5:265, 559, 1257–58; 6:1187.

72. *Naval Documents*, 5:813–19.

73. "Letter of Thomas Cushing to Robert Treat Paine, September 9, 1776," *Naval Documents*, 6:755.

74. *Journals of Congress*, 3:280.

75. For a concise account of attempts to blockade the Hudson, see Benjamin Franklin Fackenthal, Jr., "The Great Chain at West Point," *Proceedings of the Bucks County Historical Society*, 7 (1937):569–611.

In 1776, Congress authorized the Pennsylvania Committee of Safety to construct a similar boom on the Delaware River opposite Billingsport, New Jersey. There, however, the boom was considered the major navigational obstruction (*Journals of Congress*, 5:443).

76. Ransom, *Vanishing Ironworks*, pp. 301–02, finds no clear evidence that the chain was cast at Ringwood Furnace as Fackenthal and others have claimed. See also Edward C. Boynton, *History of West Point* (New York: D. Van Nostrand, 1863), pp. 39–47.

77. Ransom, *Vanishing Ironworks*, pp. 303–15; Boynton, *West Point*, pp. 48–78.

78. Aitcheson, *History of Metals*, 2:317–18.

79. James Russell Trumbull, *History of Northampton, Massachusetts*, 2 vols. (Northampton: Gazette Printing Co., 1898), 1:358–67.

80. *N.Y. Records*, 6:127; 8:449. Peter Hasenclever claimed to have been given a one-sixth share in this "very valuable" mine (*Peter Hasenclever, merchant*, pp. 25–26).

81. Washington, *Writings*, 3:420, 424.

82. James Sullivan, ed., *Minutes of the Albany Committee of Correspondence, 1775–1778*, 2 vols. (Albany: University of the State of New York, 1923), 1:205.

83. Philip H. Smith, *General History of Duchess County* (Pawling: By the author, 1877), p. 29.

84. *Conn. Records*, 15:37; Royal R. Hinman, ed., *A Historical Collection of the Part Sustained by Connecticut during the War of the Revolution* (Hartford: E. Gleason, 1842), pp. 264, 313.

85. Darwin H. Stapleton, "General Daniel Roberdeaux and the Lead Mine Expedition, 1778–1779," *Pennsylvania History*, 38 (1971):361–71; *Report of the Commission to Locate the Site of the Frontier Forts of Pennsylvania*, 2 vols. (Harrisburg Clarence M. Bush, 1896), 1:499–503.

86. William E. Pulsifer, *Notes for a History of Lead* (New York: D. Van Nostrand, 1888), p. 76; *N.C. Records*, 10:565, 598, 927, 956.

87. Henry Rowe Schoolcraft, *A View of the Lead Mines of Missouri* (New York: Charles Wiley & Co., 1819), pp. 9–20; Ruby Johnson Swartzlow, "The Early History of Lead Mining in Missouri," *Missouri Historical Review*, 28 (1934):109–14, 194–204; 29 (1934–35):27–34, 184–94, 287–95.

88. *Penn. Records*, 10:558, 637; Sullivan, ed., *Albany Committee*, 1:812.

89. Peter Force, ed., *American Archives*, 5th series, 3 vols. (Washington, D.C.: M. St. Clair and Peter Force, 1848–53), 1:144.

90. A. J. Wall, "The Statue of King George III and the Honorable William Pitt Erected in New York City, 1770," *New York Historical Society Quarterly*, 4 (1920):44–54. Fragments of the statue have recently been found in a swamp near Wilton, Connecticut (*The New York Times*, February 22, 1973, p. 43).

91. David John Mays, *Edmund Pendleton, 1721–1803*, 2 vols. (Cambridge: Harvard University Press, 1952), 1:203–05.

92. David John Mays, ed., *The Letters and Papers of Edmund Pendleton*, 2 vols. (Charlottesville: University of Virginia Press, 1967), 1:154.

93. Thomas Jefferson, *The Papers of Thomas Jefferson*, Julian P. Boyd, ed., 19 vols. (Princeton: Princeton University Press, 1958–71), 1:460.

94. W.W. Henning, ed., *The Statutes at Large, Being a Collection of All the Laws of Virginia, 1619–1792*, 13 vols. (Richmond: J. and G. Cochran, 1821), 9:237–38.

95. Jefferson, *Papers*, 2:350.

96. Ibid., 3:325, 479, 480. For a description of the Tory uprising of 1780, which had as one of its principal objectives the capture of the lead mines, see Louise Phelps Kellogg, ed. *Frontier Retreat on the Upper Ohio, 1779–1781* (Madison: Wisconsin Historical Society, 1917), 24:23–28, 195 ff.

97. Donald E. Reynolds, "Ammunition Supply in Revolutionary Virginia," *Virginia Magazine of History and Biography*, 73 (1965):64–77.

98. Paul Leicester Ford, ed., *The Writings of Thomas Jefferson*, 10 vols. (New York: G. P. Putnam's Sons, 1892–99), 3:111–13. In the same passage in the

Notes, Jefferson also noted the existence of the rich lead fields around Kaskaskia on the Mississippi. Considering Jefferson's strenuous efforts to provide lead for the colonial armies, it is interesting to speculate to what extent the presence of ample supplies of lead and the anticipation of other metallic wealth from the trans-Mississippi region influenced his judgment in making the Louisiana Purchase.

99. Ibid., 5:191–92, 199, 263, 367.

100. Ibid., 5:600.

101. "Accounts of Col. John Lynch, Superintendent of the Mines, Annual Reports, 1779–1782," in the Archival Collections, Virginia State Library, Richmond.

102. A prime source of records for the trade of colonial craftsmen, the colonial newspapers, breaks down during the Revolution when publication was intermittent. There is little question, however, that the practice of the crafts underwent severe disarrangement.

103. Thompson Westcott, *The Life of John Fitch* (Philadelphia: J. B. Lippincott, 1878), pp. 63–77.

104. Bruff found it expedient to move to Nova Scotia after the British evacuated New York (Gottesman, ed., *Arts and Crafts in New York,* pp. 63–64).

105. Washington, *Writings,* 10:278, 280, 330.

106. "Public Foundry Journal, Feb. 1, 1779 to June 19, 1781," [Virginia], pp. 2, 24, in the Archival Collections, Virginia State Library, Richmond.

107. "Account of work performed by Carpenters, Smiths, etc., at the Armoury, Jan. 1 to Aug. 31, 1780," Record Group 93, Roll 72, item 21806.

108. *Journals of Congress,* 7:193, 228, 272.

109. Ibid., 7:324. Also, "Letter from Joseph Belton to the Continental Congress, April 4, 1777," Record Group 360, Roll 41, 1, fol. 123.

110. *Penn. Archives,* 1st series, 4:712,767–68.

111. Ibid., p. 717.

112. Rhode Island stipulated a price of £3 18s. for firearm, bayonet, ramrod, and cartouche box in March 1776. See *R.I. Records,* 7:477–78; *N.C. Records,* 10:981.

113. "Baxter Manuscripts," *Collections of the Maine Historical Society,* 2d series, 24 vols. (Portland, 1869–1916), 14:443–44.

114. *Journals of Congress,* 12:1059. The full proposal is in Record Group 360, Roll 41, 8, fol. 60.

115. "Letter from M. Penet to Congress, May 20, 1778," Record Group 360, Roll 147, 6, fol. 231.

Chapter 8

1. Alexander Hamilton, *Report on Manufactures,* December 5, 1791, Hamilton, *Papers;* the full text is contained in 10:230–340.

2. For the most complete exposition of the mercantile system see Adam Smith, *The Wealth of Nations,* 2 vols. (1776; Oxford: Clarendon Press, 1880), chap. 8.

3. Ibid., 2:165.

4. John Lord Sheffield, *Observations on the Commerce of the American States,* "New Edition" (1784; reprinted, New York: Augustus M. Kelley, 1970), pp. 191–93.

5. Ibid., pp. 5–7, 13–21, 29–30, 31. In the case of iron, the production of axes, scythes, and iron castings were the only possible exceptions noted.

6. Under the Articles of Confederation, the individual states retained the rights to establish tariffs on imports and to regulate prices within their boundaries, so long as such practice did not conflict with the stipulations of any international treaties. No treaty of commerce had been enacted between Sweden and the United States at the time of Hermelin's report, *Mines of the United States,* pp. 9–14.

7. Ibid., pp. 46, 49–50, 60, 74.

8. Ibid., pp. 61–63.

9. Ibid., pp. 67–69. Monthly price index tables for domestic and foreign iron and other metals from 1784 onward are listed in Anne Bezanson, Robert D. Gray, and Miriam Hussey, *Wholesale Prices in Philadelphia, 1784–1861,* Part I, discussion, Part II, tables (Philadelphia: University of Pennsylvania Press, 1936–37). Cf. 2:103, 104, 107. Swedish bar iron prices averaged $1 to $3 per ton below domestic during the 1780s and 1790s.

10. The estimate of parity is based on the calculations of Bining, *British Regulations,* pp. 24–30. In the absence of exact statistics, Bining offers the best available comparison of the two industries for that period.

11. Ibid., p.118.

12. Hermelin, *Mines of the United States,* pp. 69–76.

13. A discussion of the fluctuations of the price of pig and bar iron in Philadelphia from 1770 to 1790 is found in Anne Bezanson, *Prices and Inflation during the American Revolution* (Philadelphia: University of Pennsylvania Press, 1951), pp. 149–74.

14. See John Spargo, *Iron Mining and Smelting in Bennington, Vermont, 1786–1842* (Bennington: Bennington Historical Museum, 1938); Ralph W. Putnam, "Vermont's Part in Industry," *Vermont Historical Society Proceedings,* New Series, 8 (1940):357–59.

15. Wallace, *South Carolina,* pp. 454–55.

16. Scrivenor, *Iron Trade,* pp. 86–89.

17. 25 George III, c. 67; Pickering, *Statutes,* 35:286–91.

18. Walker, "The End of Colonialism in the Middle Atlantic Iron Industry," pp. 5–26.

19. Tench Coxe, "Statements relative to the agriculture, manufactures, commerce, population, resources and public happiness of the United States, in reply to the assertions and predictions of Lord Sheffield," in *A View of the United States of America* (Philadelphia: William Hall, 1794), pp. 144–46.

20. The development of a tariff policy for the United States entailed the interaction of very complex factors, and only the major trends can be summarized here. For a full analysis see William Hill, "The First Stages of the Tariff Policy of the United States," *Proceedings of the American Economic Association,* 8 (1893):452–614, on which the following discussion is based.

21. Smith, *Wealth of Nations,* 2:162.

22. William Wirt, *The Life of Patrick Henry* (New York: McElrath & Bangs, 1831), p. 255.

23. An example of Deane's jeremiads is included in James Ferguson, ed., *The*

Papers of Robert Morris, 1781–1784 (Pittsburgh: University of Pittsburgh Press, 1973), pp. 130 ff.

24. "John Adams to John Luzac, September 15, 1780," Adams, *Works,* 8:255.

25. Thomas Jefferson, "Notes on the State of Virginia," Ford, ed., *Writings of Thomas Jefferson,* 3:269. Jefferson maintained this position for years, but in 1816 regretfully admitted, "Experience has taught me that manufactures are now as necessary [as agriculture] to our independence as to our comfort" (ibid., 10:10).

26. Alexander Hamilton, "A Full Vindication of the Measures of Congress, from the Calumnies of their Enemies," Hamilton, *Papers,* 1:63.

27. Edward Stanwood, *American Tariff Controversies in the Nineteenth Century,* 2 vols. (Boston: Houghton Mifflin, 1904), 1:23–24. A principal stumbling block to state acceptance of the tariff acts proposed by Congress was not the issue of free trade but fear of encroachment by the federal government on the jealously guarded rights to tax guaranteed to the states by the Articles.

28. In the decade from 1780 to 1790, Virginia passed more tariff legislation than any other state but only for the purpose of raising revenue. See Henning, *Statutes at Large,* 10:150, 165, 281, 382; 11:95, 121, 201, 374; 12:32, 46, 288, 290, 304, 412, 442, 514.

29. William Wirt Henry, *The Life, Correspondence and Speeches of Patrick Henry,* 3 vols. (New York: Charles Scribner's Sons, 1891), 2:328–29.

30. Hill, "Tariff Policy," p. 503.

31. Albert Stillman Batchellor et al., eds., *Laws of New Hampshire, 1679–1835,* 10 vols. (Manchester and Concord: 1904–22), 5:146–48.

32. James T. Mitchell and Henry Flanders, comps., *The Statutes at Large of Pennsylvania from 1682 to 1801,* 16 vols. (Harrisburg: 1896–1911), 12:99–104.

33. "To the Honourable Representatives of the Freeman of the Commonwealth of Pennsylvania, in general Assembly met. The Petition of the Subscribers, Manufacturers of Bar Iron within the Commonwealth aforesaid, November 30, 1785," Frank B. Nead Documents, 1663–1866, Collections of the Historical Society of Pennsylvania, Philadelphia.

34. "James Madison to Edmund Pendleton, January 9, 1787," James Madison, *Letters and Other Writings of James Madison,* 4 vols. (New York: R. Worthington, 1884), 1:271.

35. Hill, "Tariff Policy," p. 496.

36. *Constitution of the United States,* Article I, Section X: "No state shall, without the consent of Congress, lay any imposts or duties on imports and exports, except what may be absolutely necessary for executing its inspection laws; and the net produce of all duties and imposts, laid by any State on imports and exports, shall be for the use of the Treasury of the United States, and all such laws shall be subject to the revision and control of the Congress."

37. Joseph Gales, comp., *The Debates and Proceedings in the Congress of the United States,* vol. I: *March 3, 1789, to March 3, 1791* (Washington, D.C.: Gales and Seaton, 1834), pp. 106–07.

38. Ibid., pp. 110–11.

39. Walter Lowrie et al., eds., *American State Papers, Finance,* 5 vols. (Washington, D.C.: 1832–58), 1:5–11. No breakdown by category of goods is available, but the total value of British exports to and imports from the United States during the 1780s in pounds sterling were officially listed as:

Year	Imports	Exports
1784	749,345	3,679,467
1785	893,594	2,308,023
1786	843,119	1,603,465
1787	893,637	2,009,111
1788	1,023,789	1,886,142
1789	1,010,198	2,525,298
1790	1,191,071	3,431,778

See Timothy Pitkin, A *Statistical View of the Commerce of the United States* (Hartford: Charles Hosmer, 1816), p. 30.

40. Gales, comp., *Debates in Congress*, pp. 156–57.

41. *Tariff Acts Passed by the Congress of the United States from 1787 to 1909* (Washington, D.C.: 1909), pp. 13–15.

42. Hamilton, *Papers*, 10:230 ff. The merits and intent of Hamilton's *Report* and Tench Coxe's role in its compilation recently have been reevaluated by John R. Nelson, Jr., "Alexander Hamilton and American Manufacturing: A Reexamination," *The Journal of American History*, 65 (1979):971–95, and Jacob E. Cooke, "Tench Coxe, Alexander Hamilton, and the Encouragement of American Manufactures," *The William and Mary Quarterly*, 3d series, 32 (1975):369–92.

43. Although Hamilton directed his opposition mainly to the agrarian arguments propounded by Adam Smith in *The Wealth of Nations*, many of Smith's statements were similar to those used by Thomas Jefferson and other advocates of a free-trade, agrarian America.

44. Hamilton, *Papers*, 10:246 (italics in original).

45. Ibid., 10:314–19.

46. More, *Utopia*, p. 43.

47. Hamilton, *Papers*, 10:314.

48. Ibid., 9:335.

49. Ransom, *Vanishing Ironworks*, pp. 162–63.

50. Coxe, "Statements," p. 144.

51. Pierce, *Iron in the Pines*, pp. 16–17, 71, 84–86.

52. Bining, *Pennsylvania Iron*, Appendix A, pp. 189–92.

53. For a more complete discussion of the cementation and crucible processes for making steel, see Aitcheson, *History of Metals*, 2:453–59.

54. Nearly a third of the furnaces listed in the 1810 census still were "air" furnaces in which no blast was used to smelt ore.

55. A copy of the royal proclamation announcing the ban on West Indian trade is found in Adams, *Works*, 8:97–98 n.

56. Arthur Cecil Bining, "The Rise of Iron Manufacture in Western Pennsylvania," *Western Pennsylvania Historical Magazine*, 16 (1933):234–56; Wilbur Stout, "Early Forges in Ohio," *Ohio Archaeological and Historical Quarterly*, 66 (1937):25–41.

57. Henry C. Grittinger, "The Iron Industries of Lebanon County," *Papers and Addresses of the Lebanon County Historical Society*, 3 (1904):10.

58. The effect of Henry Cort's puddling and rolling patents on the coke smelting of iron in Great Britain is discussed by Theodore S. Wertime, *The Coming of the Age of Steel* (Chicago: The University of Chicago Press, 1962), pp. 126–28, 229–34.

59. For an economic analysis of the relative merits of charcoal versus coke smelting in the early American iron industry see Peter Temin, *Iron and Steel in Nineteenth-Century America* (Cambridge: The M.I.T. Press, 1964), chap. 3.

60. Raymond F. Hunt, Jr., "The Pactolus Ironworks," *Tennessee Historical Quarterly*, 25 (1966):176–96.

61. J. Peter Lesley, *The Iron Manufacturers' Guide to the Furnaces, Forges and Rolling Mills of the United States* (New York: John Wiley, 1866), p. 126.

62. Evelyn Abraham, "Isaac Meason, the First Ironmaster West of the Alleghenies," *Western Pennsylvania Historical Magazine*, 20 (1937):41–49.

63. *American State Papers, Finance*, 2:695–96.

64. Pitkin, *Commerce of the United States*, pp. 62–63.

65. Bezanson, Gray, and Hussey, *Wholesale Prices*, 2:54.

66. Ibid., pp. 47–52. The practice of sheathing ship bottoms with copper to prevent attack by marine organisms began in England about 1761 and was fully established by 1790. John Raymond Harris, "Copper and Shipping in the Eighteenth Century," *Economic History Review*, Series 2, 19 (1966):550–68, recounts the development of ship sheathing in Great Britain. No comparable study exists for the United States. Sheathing may have been used as early as 1784, but it did not become common practice before the end of the century. Cf. Maxwell Whitman, *Copper for America* (New Brunswick: Rutgers University Press, 1971), pp. 249–50.

67. J. H. Granbery, "The Schuyler Mine," *Journal of the Franklin Institute*, 164 (1907):23–24; B. Henry Latrobe, *American Copper Mines* (Report on Schuyler Mine made to N.J. Legislature's Committee of Commerce and Manufactures) (n.p., n.d. [ca. 1801]).

68. Epaphroditus Peck, *A History of Bristol, Connecticut* (Hartford: The Lewis Street Bookshop, 1932), pp. 134–35.

69. The works produced the roofing sheet for the Virginia State House ("Edward Carrington to Hamilton, October 4, 1791," Hamilton, *Papers*, 9:277).

70. Joseph Schaefer, *The Wisconsin Lead Region* (Madison: State Historical Society of Wisconsin, 1952), p. 253.

71. Miriam Hussey, *From Merchants to Colour Men* (Philadelphia: University of Pennsylvania Press, 1956), pp. 18–24, 130–31.

Chapter 9

1. Anonymous pamphlet, "Reflections on the Importation of Bar-Iron, from our own Colonies of North America in Answer to a late Pamphlet on the Subject" (London, 1757), p. 3, in the Archival Collections, Library of Congress, Washington, D.C.

2. Until recently, the works of Arthur Cecil Bining on colonial iron manufactures and Henry Kauffman's treatment of colonial crafts were among the few exceptions to this statement. Standard college textbooks dealing with colonial history still make scant reference to iron manufacture or metals crafts during the period.

3. *Mass. Records*, 2:61–62.

4. In charcoal smelting, even with high-grade ores, yields were low. Much iron was lost with the slag. Peter Hasenclever's introduction of a stamping mill to crush furnace cinder was an effort to recover that metal. See report of the "Franklin Commission," *N.J. Archives*, 1st series, 28:252. The small bodies of high-grade copper ore in the East, often containing up to 99 percent pure native copper, were soon exhausted, most before the end of the colonial period.

5. Arber and Bradley, eds., *John Smith*, 1:12–13.

6. After an iron industry was established, many of the tools could be manufactured in America, but specialty items, like bronze molds for cast pewterware and steel-cutting tools, still had to be imported from England.

7. "Value and Quantity of Articles Exported from British Continental Colonies, By Destination: 1770," *Historical Statistics*, p. 761.

8. See Table 1.

9. Ibid.

10. Green, "Gold Mining," pp. 3–7. A more complete account of the Conrad Reed story is given in Bruce Roberts, *The Carolina Gold Rush* (Charlotte: McNally and Loftin, 1971), pp. 5–8.

Bibliographic Essay

The archival records and published literature relating to the history of metals in colonial America are extensive. Archival material is widely dispersed in state and local history collections from New Hampshire to South Carolina. Although many of the smaller collections may afford opportunities for further study, the major collections on iron manufacture of the Historical Society of Pennsylvania and records of the Continental Congress and War Department in the National Archives illustrating the use of metals in the Revolutionary and Confederation periods offer the greatest potential for future research in this field.

No study of metals in colonial America would be complete without a consideration of the thousands of artisans who comprised the majority of metalworkers during the period. Outstanding collections of artifacts exist in a number of museums, together with records on many of the craftsmen. In particular, the collections and records of the Smithsonian Institution in Washington, D.C., the Henry Francis du Pont Winterthur Museum in Delaware, the Museum of Decorative Arts in Winston-Salem, North Carolina, the Boston Museum of Fine Arts and Old Sturbridge Village in Massachusetts, Colonial Williamsburg, Virginia, and the Bucks County Historical Society in Pennsylvania are among the most impressive. For a more comprehensive listing see Lucius F. Ellsworth, "A Directory of Artifact Collections," in Brooke Hindle, *Technology in Early America* (Chapel Hill, 1966).

Many books and articles are specifically devoted to some aspect of the manufacture of metals during the colonial period. Within the last three decades, these have begun to appear in greater numbers and to exhibit a higher level of scholarship than earlier works, although many of the latter still retain value to a student of the subject. In addition, the great number of state and local histories, biographies, and general histories that pertain to the colonial period often contain some reference to metals. Included in this category are the many articles published annually in the journals of state and local historical societies. Throughout the present study, I attempted to review as many secondary publications as possible for such references, and although the survey was not all-inclusive, much valuable information was obtained, as the footnotes attest.

Secondary works, however, are never completely satisfactory in reconstructing the historical record for which primary material alone will suffice. Here also a voluminous literature exists in the published records of colonial and early state governments, in the departments of the British government, and in the papers and correspondence of prominent colonial figures. One of the purposes of writing history is to present the findings of the author, while sparing the reader the ordeal of repeating the process by which these findings were determined. It is with this thought in mind that this Bibliographic Essay has been written. Only those primary and secondary sources of more than incidental value have been included. The titles cover the full spectrum of topics treated in the study and will afford the interested reader the opportunity to pursue those topics in greater depth.

Official Printed Sources

Great Britain

Public Record Office, *Calendar of State Papers, Colonial Series, Preserved in the Public Record Office*, 38 vols. London, 1860–1919.

Proceedings and Debates of the British Parliaments Respecting North America. Leo Francis Stock, ed. Vols I–V. Washington, D.C., 1924–41.

Statutes at Large, from Magna Charta to 1807. Danby Pickering et al., comps. 46 vols. Cambridge and London, 1762–1808.

United States

American Archives. Peter Force, ed. Fourth Series, 6 vols. Fifth Series, 3 vols. Washington, D.C., 1837–46.

Journals of the Continental Congress, 1774–1789. Worthington C. Ford, et al., eds. 34 vols. Washington, D.C., 1904–37.

Historical Statistics of the United States: Colonial Times to 1957. Washington, D.C.: U.S. Department of Commerce, 1960.

Naval Documents of the American Revolution. 7 vols. Washington, D.C., 1964–76.

Tracts and Other Papers, Relating Principally to the Origin, Settlement, and Progress of the Colonies in North America. Peter Force, ed. 4 vols. Washington, D.C., 1836–46.

Public Records of the Colony of Connecticut, 1636–1776. J. H. Trumbull and C. J. Hoadley, comps. 15 vols. Hartford, 1850–90.

Public Records of the State of Connecticut. C. J. Hoadley et al., comps. 9 vols. Hartford, 1894–1953.

New Haven Colonial Records. C. J. Hoadley, comp. 2 vols. Hartford, 1857–58.

Archives of Maryland. W. H. Browne et al., eds. 70 vols. Baltimore, 1883–1964.

Records of the Governor and Company of the Massachusetts Bay in New England, 1628–1686. Nathaniel B. Shurtleff, ed. 5 vols. Boston, 1853–54.

Archives of the State of New Jersey. W. A. Whitehead et al., eds. First Series, Documents Relating to the Colonial History, 1631–1800. 42 vols. Newark, etc., 1880–1949.

Archives of the State of New Jersey. J. L. Murphy, ed. Second Series, 5 vols. Trenton, 1901–17.

Documents Relative to the Colonial History of the State of New York. E. B. O'Callaghan and B. Fernow, eds. 15 vols. Albany, 1856–87.

Colonial and State Records of North Carolina, 1662–1790. William L. Saunders and Walter Clark, eds. 26 vols. Raleigh, Winston, etc., 1886–1906.

Pennsylvania Colonial Records, 1683–1790. 16 vols. Vols. I–X: *Minutes of the Provincial Council;* Vols XI–XVI: *Minutes of the Supreme Executive Council.* Philadelphia, 1860.

Pennsylvania Archives, 1st series, Samuel Hazard, ed. 12 vols. Philadelphia, 1852–56.

Records of the Colony of Rhode Island and Providence Plantations in New England, 1636–1792. J. R. Bartlett, comp. 10 vols. Providence, 1856–65.

General Works

A number of nineteenth-century works dealt wholly or in part with colonial metals manufactures. In particular, the several editions of J. Leander Bishop, *A History of American Manufactures from 1608 to 1860* (3 vols., Philadelphia: E. Young, 1868), and James H. Swank, *A History of the Manufacture of Iron in All Ages* (Philadelphia, 1884), have served as the starting point for many later studies, such as those of Victor S. Clark, *History of Manufactures in the United States* (3 vols., New York, 1929). Although these works contain material difficult to find elsewhere, it is not always easy to verify, and they must be used with caution.

On the British side, H. R. Schubert, *History of the British Iron and Steel Industry from c. 450 B.C. to A.D. 1795* (London: Routledge & Kegan Paul, 1957), and T. S. Ashton, *Iron and Steel in the Industrial Revolution* (Manchester: Manchester University Press, 1951), are standard sources for the iron industry. Harry Scrivenor, *History of the Iron Trade* (2d ed., London: Longman, Brown, Green and Longmans, 1854), although dated, is still useful. Henry Hamilton, *The English Brass and Copper Industries to 1800* (London, 1967), presents the story for the major nonferrous metals of the period.

Chapter 1

The literature on the voyages of exploration is extensive although attention to metals often is limited. Samuel Eliot Morison, *The European Discovery of America: The Northern Voyages, A.D. 500–1600* (New York: Oxford University Press, 1971), is an excellent modern summary. Contemporary accounts of many early voyages are contained in Richard Hakluyt, *The Principal Navigations, Voyages, Traffiques and Discoveries of the English Nation* (10 vols., New York: E. P. Dutton, 1927–28). Two books giving particular attention to metals are David Beers Quinn, ed., *The Voyages and Colonizing Enterprizes of Sir Humphrey Gilbert* (2 vols., London: Hakluyt Society, 1940), and Vilhjalmur Stefansson, *The Three Voyages of Martin Frobisher* (2 vols., London: Argonaut Press, 1938), the latter containing a discussion of the "ore" found at Hall's Island. Pre-Columbian mining and metallurgy are discussed by Dudley T. Easby, Jr., "Pre-hispanic Metallurgy and Metalworking in the New World," *Proceedings of the American Philosophical Society*, 109 (1963):89–98.

Chapter 2

Attitudes toward metals among the earliest Jamestown colonists may be found in the contemporary accounts included in Edward Arber and A. G. Bradley, eds., *The Travels and Works of Captain John Smith* (Edinburgh: J. Grant, 1910). Louis B. Wright, *The Dream of Prosperity in Colonial America* (New York: New York University Press, 1965), stresses the importance of metals in the thinking of the settlers. Wesley Frank Craven, *Dissolution of the Virginia Company* (Gloucester: Peter Smith, 1964), argues for a more restrained interpretation. The Falling Creek iron furnace venture is documented by Charles E. Hatch, Jr., and Thurlow Gates Gregory, "The First American Blast Furnace, 1619–1622," *The Virginia Magazine of History and Biography*, 70 (1962):259–96. Edward Neal Hartley, *Ironworks on the Saugus* (Norman:

University of Oklahoma Press, 1957), is the standard history of the Hammersmith operation.

Chapter 3

Prehistoric copper mining in North America is described in Roy Ward Drier and Octave Joseph Du Temple, eds., *Prehistoric Copper Mining in the Lake Superior Region: A Collection of Reference Articles* (Calumet: Privately published, 1961), and George Brinton Phillips, "The Primitive Copper Industry of America," Part I, *Journal of the Institute of Metals*, 34 (1925):261–67; Part II, 36 (1926):99–106. A geological discussion of colonial copper mining in New Jersey is the treatment of Herbert P. Woodward, *Copper Mines and Mining in New Jersey*, Bull. 57, Geologic Series, Department of Conservation and Development, State of New Jersey (Trenton, 1944). Harry B. Weiss and Grace M. Weiss, *Old Copper Mines of New Jersey* (Trenton: Past Times Press, 1963), gives a general overview. The operations of the important Schuyler mine are discussed by Elizabeth Marting, "Arent Schuyler and His Copper Mine," *Proceedings of the New Jersey Historical Society*, 65 (1947):126–40, and J. H. Granbery, "The Schuyler Mine," *Journal of the Franklin Institute*, 164 (1907):13–28. The Simsbury mine in Connecticut is one of the few colonial copper mines outside of New Jersey that has been fully documented; see Richard H. Phelps, *Newgate of Connecticut: Its Origin and Early History* (Hartford: American Publishing Co., 1876). Elsewhere, George F. Dow, "The Topsfield Copper Mines," *Proceedings of the Massachusetts Historical Society*, 65 (1936):570–80, describes a brief attempt to mine copper in Massachusetts, and the writings of William Byrd of Westover, particularly *A Journey to the Land of Eden and Other Papers*, Mark Van Doren, ed. (New York: Macy-Masius, 1928), provide a contemporary account of copper diggings in southeastern Virginia.

Chapter 4

For iron manufacture in the eighteenth century, Arthur Cecil Bining, *Pennsylvania Iron Manufacture in the Eighteenth Century* (Harrisburg: Pennsylvania Historical and Museum Commission, 1938), remains unsurpassed for its coverage of the technology. A recent work by W. David Lewis, *Iron and Steel in America* (Greenville, 1976), provides a more general overview of the period. Older works containing some useful information on local "firsts," furnaces or forges, are James H. Swank, and J. B. Pearse, *Concise History of the Iron Manufacture of the American Colonies up to the Revolution and of Pennsylvania until the Present Time* (Philadelphia, 1876). Beyond Hartley's work on Saugus, no general account of iron smelting in eighteenth-century Massachusetts has been written, one of the more serious omissions in the historical record. Similarly, the fine short article by Irene D. Neu, "The Iron Plantations of Colonial New York," *New York History*, 33 (1952):3–24, is the most important work on that state's industry, which awaits further documentation. Herbert C. Keith and Charles Rufous Harte, "The Early Iron Industry of Connecticut," *Fifty-first Annual Report of the Connecticut Society of Civil Engineers* (1935), is the only full account for that colony. For New Jersey, a number of good histories exist, beginning with Charles E. Boyer, *Early Forges and Furnaces in New Jersey* (Philadelphia: University of Pennsylvania Press, 1931). More regionally oriented are the works of Arthur D. Pierce, *Iron in the Pines* (New

Brunswick: Rutgers University Press, 1957), on the south central industry, and James M. Ransom, *Vanishing Ironworks of the Ramapos* (New Brunswick: Rutgers University Press, 1966), covering the New Jersey–New York border region.

In Maryland, Henry Whitley's "The Principio Company," *The Pennsylvania Magazine of History and Biography*, 11 (1887):63–68, 190–98, 289–95, remains the major published account of that early operation although the doctoral dissertation of Michael Warren Robbins, "The Principio Company: Iron-Making in Colonial Maryland, 1720-1821" (The George Washington University, 1972), surpasses it in breadth and depth of treatment. Keach Johnson's articles, "The Genesis of the Baltimore Ironworks," *Journal of Southern History*, 19 (1953):157–80, and "The Baltimore Company Seeks English Markets: A Study of the Anglo-American Iron Trade, 1731–1755," *William and Mary Quarterly*, 3rd series, 16 (1959):37–60, provide a picture of the workings of that important industry. For Virginia, Robert Alonzo Brock, "Early Iron Manufacture in Virginia: 1619–1776," *Proceedings of the United States National Museum*, 8 (Washington, 1886), has never been superseded although the opening chapters of Kathleen Bruce, *Virginia Iron Manufacture in the Slave Era* (New York: The Century Co., 1930), contain additional information on the period. Again, William Byrd provides perhaps the finest contemporary account of an iron furnace in Virginia, in *A Progress to the Mines*, in *The Prose Works of William Byrd of Westover*, Louis B. Wright, ed. (Cambridge: Harvard University Press, 1966).

Chapter 5

The literature on colonial artisans, including metalworkers, probably comprises the largest volume of history on any subject covered by this study. For a single work putting the metals crafts in the perspective of all colonial manufacture, Carl Bridenbaugh's *The Colonial Craftsman* (New York: New York University Press, 1950) is the best available. Edwin Tunis, *Colonial Craftsmen and the Beginning of American Industry* (Cleveland, 1965), offers extensive illustrations of tools and techniques. Rita Gottesman, ed., *The Arts and Crafts in New York, 1726–1776* (New York: New York Historical Society, 1938), and George Francis Dow, *The Arts and Crafts in New England, 1705–1775* (Topsfield: The Wayside Press, 1927), look at the products of colonial craftsmen through an analysis of contemporary newspaper advertisements. The works of Henry J. Kauffman cover individual crafts in some depth: *Early American Ironware* (Rutland: Charles E. Tuttle, 1966), *American Copper and Brass* (Camden: Thomas Nelson & Sons, 1968), *The Colonial Silversmith* (Camden: Thomas Nelson & Sons, 1969), *The American Pewterer: His Techniques and His Products* (Camden: Thomas Nelson & Sons, 1970).

Material on blacksmithing is limited considering the importance of this craft to colonial society, although Albert H. Sonn, *Early American Wrought Iron* (3 vols., New York: Charles Scribner's Sons, 1928), offers profuse illustrations for one phase of the ironworker's trade. Joseph A. Goldenberg, *Shipbuilding in Colonial America* (Charlottesville: University Press of Virginia, 1976), discusses the general state of shipbuilding in the period, but too little is known about colonial shipwrights and allied craftsmen. The same is largely true about colonial coppersmiths, although one specialized phase of the trade is well covered by Silvio A. Bedini, *Early American Scientific Instruments and Their*

Makers (Washington, D.C.: United States National Museum Bulletin 231, 1964). Two fine biographies of men who were accomplished coppersmiths although better known for other contributions to colonial society are Esther Forbes, *Paul Revere and the World He Lived In* (Boston: Houghton Mifflin, 1944), and Brooke Hindle, *David Rittenhouse* (Princeton: Princeton University Press, 1964).

American silver and pewter have been well researched; in the case of silver, one or more books exist for the practice in each of the original states. Martha Gandy Fales, *Early American Silver* (New York: E. P. Dutton, 1973), and Graham Hood, *American Silver: A History of Style, 1650–1900* (New York: Praeger Publishers, 1971), are sound general works with excellent illustrations and adequate bibliographies. George Barton Cutten's works on *The Silversmiths of North Carolina* (Raleigh: State Department of Archives and History, 1948) and *Silversmiths of Virginia* (Richmond: The Dietz Press, 1952) are good examples of the regional history genre. Hermann Frederick Clarke, "John Hull: Mintmaster," *New England Quarterly*, 10 (1937):668–84, is an interesting sketch of one of the earliest major silversmiths. Ledlie I. Laughlin, *Pewter in America* (2 vols., Boston: Houghton Mifflin, 1940), remains the standard work. Charles F. Montgomery, *A History of American Pewter* (New York, 1973), takes full advantage of recent scholarship to supplement Laughlin. Shirley Spaulding DeVoe, *The Tinsmiths of Connecticut* (Middletown: Wesleyan University Press, 1918), still is the only major work devoted to the early history of tin manufacture in America.

Chapter 6

The political struggle between the American colonies and Great Britain in the eighteenth century has been well documented by historians. The contribution of increasing manufacturing to the growing dissension is the subject of Arthur Cecil Bining's excellent study, *British Regulation of the Colonial Iron Industry* (Philadelphia: University of Pennsylvania Press, 1933). An excellent cross-section of many British and American pamphlets generated by the debates over regulating iron is in the Manuscript Division of the Library of Congress. Bining cites a number of these as well as protests in the writings of many statesmen on both sides of the Atlantic. The prerevolutionary activities of Peter Hasenclever are fully recounted in his self-serving volume, *The Remarkable Case of Peter Hasenclever, merchant* (London, 1773). A more balanced, and available, account is Gerhard Spieler, "Peter Hasenclever, Industrialist," *Proceedings of the New Jersey Historical Society*, 59 (1941):231–56.

Chapter 7

The material in this chapter largely is gathered from a broad spectrum of official published records and papers and from archival sources. Detailed studies of metals and the Revolution are few. The several excellent military histories of the revolutionary war, such as John Richard Alden, *A History of the American Revolution* (New York, 1972), have little to say about the problems of ordnance procurement, the area where the colonial iron industry made a major contribution to the war effort. The stores of information contained in the National Archives Record Group 360 (Records of the Continental and Confederation Congresses) and 93 (War Department Collection of Revolutionary War Records) have yet to be adequately digested by historians. Case studies such as Thomas M. Doerflinger, "Hibernia Furnace during the Revolution," *New Jersey History*, 90 (1972):97–114, remain to be written for other colonial ironworks. Boris Erich Nelson, "New Jersey Iron: The State Museum Exhibit," *Proceedings of the New Jersey Historical Society*, 72 (1954):270–73, notes the role of that state's iron industry in the Revolution.

Two early articles, M. V. Brewington, "American Naval Guns, 1775–1785," *The American Neptune*, 3 (1943):11–18, 148–58, and B. F. Fackenthal, Jr., "The Great Chain at West Point," *Proceedings of the Bucks County Historical Society*, 7 (1937):596–611, deal with limited topics of interest during this period, as does Darwin H. Stapleton, "General Daniel Roberdeaux and the Lead Mine Expedition, 1778–1779," *Pennsylvania History*, 38 (1971):361–71. The period has great potential for further investigation.

Chapter 8

Samuel Gustav Hermelin's postwar tour of the central Atlantic states produced the best contemporary survey of the American iron industry in his *Report about the Mines of the United States of America, 1783*, Amandus Johnson, trans. (Philadelphia: The John Morton Memorial Museum, 1931). His sober appraisal is in marked contrast to the optimism of Tench Coxe, "Statements relative to the agriculture, manufactures, commerce, population, resources and public happiness of the United States, in reply to the assertions and predictions of Lord Sheffield," in *A View of the United States of America* (Philadelphia: William Hall, 1794). For the tariff debates of the 1780s see William Hill, "The First Stages of the Tariff Policy of the United States," *Proceedings of the American Economic Association*, 8 (1893):452–614. The economic factors inhibiting the introduction of new technology in the iron industry are discussed in the opening chapters of Peter Temin, *Iron and Steel in Nineteenth-Century America* (Cambridge: The M.I.T. Press, 1964). Joseph Walker argues that English dominance in iron extended well into the nineteenth century, "The End of Colonialism in the Middle Atlantic Iron Industry," *Pennsylvania History*, 41 (1974):5–26. The drafts of Hamilton's *Report on Manufactures* are contained in Vol. 10 of *The Papers of Alexander Hamilton*, Harold C. Syrett, ed. (26 vols., New York: Columbia University Press, 1961–). Coxe's contribution to the report is argued by Jacob E. Cooke, "Tench Coxe, Alexander Hamilton, and the Encouragement of American Manufactures," *The William and Mary Quarterly*, 3d series, 32 (1975):369–92, and John R. Nelson, Jr., "Alexander Hamilton and American Manufacturing: A Reexamination," *The Journal of American History*, 65 (1979):971–95.

Chapter 9

The discovery of gold in North Carolina is recounted by Bruce Roberts, *The Carolina Gold Rush* (Charlotte: McNally and Loftin, 1971).

Index

James A. Mulholland, who teaches history at North Carolina State University, has a background in chemical engineering and the history of technology and science. He received a bachelor of science degree in chemical engineering from the Massachusetts Institute of Technology, a master's degree in liberal studies with a physics major from Wesleyan University, and a doctorate in the history of technology from the University of Delaware.